Parasitology
A Global Perspective

Parasitology
A Global Perspective

Edited by

Kenneth S. Warren, M.D.
Director, Health Sciences
The Rockefeller Foundation

and

John Z. Bowers, M.D.
Consultant, Health Sciences
The Rockefeller Foundation

With Seven Illustrations

Springer-Verlag
New York Berlin Heidelberg Tokyo

Kenneth S. Warren, M.D., Director, Health Sciences, The Rockefeller Foundation, 1133 Avenue of the Americas, New York, New York 10036 U.S.A.

John Z. Bowers, M.D. Consultant, Health Sciences, The Rockefeller Foundation, 1133 Avenue of the Americas, New York, New York 10036 U.S.A.

Library of Congress Cataloging in Publication Data
Main entry under title:
Parasitology : a global perspective.

   "Proceedings of a meeting in Bellagio, Italy, April, 1982"—P.
   Bibliography: p.
   Includes index.
   1. Parasitology—Congresses.   2. Parasitic diseases—Congresses.   I. Warren, Kenneth S.
II. Bowers, John Z., 1913–   . [DNLM: 1. Parasitology—Congresses.   QX 4 P223 1982]
QL757.P33   1983   616.9′6   83-4715

Typeset by University Graphics, Inc., Atlantic Highlands, New Jersey
Printed and bound by R. R. Donnelley & Sons, Harrisonburg, Virginia

9 8 7 6 5 4 3 2 1

ISBN-13: 978-1-4612-5552-9   e-ISBN-13: 978-1-4612-5550-5
DOI: 10.1007/978-1-4612-5550-5

Contributors

Ruth Arnon, PH.D

Professor, Department of Chemical Immunology, and Head of Unit for Molecular Biology of Parasitic Diseases, The Weizmann Institute of Sciences, Rehovot, Israel.

J. H. Arundel, M.V.SC.

Reader in Veterinary Parasitology, Department of Paraclinical Sciences, Veterinary Clinical Centre, University of Melbourne, Weribee, Victoria, Australia

Sir Christopher C. Booth, M.D.

Director, Medical Research Council, Clinical Research Center, Harrow, Middlesex, England

John Z. Bowers, M.D.

Consultant, Health Sciences, The Rockefeller Foundation, New York, New York, U.S.A.

André Capron M.D.

Professeur et Directeur, Centre d'Immunologie et de Biologie Parasitaire, Institut Pasteur, Lille, France

Eli Chernin, PH.D.

Professor, Department of Tropical Public Health, Harvard School of Public Health, Boston, Massachusetts, U.S.A.

George A. M. Cross, PH.D.

Andre and Bella Meyer Professor, Molecular Parasitology, The Rockefeller University, New York, New York, U.S.A.

John R. David, M.D.
John LaPorte Given Professor and Chairman, Department of Tropical Public Health, Harvard School of Public Health, and Professor of Medicine, Harvard Medical School, Boston, Massachusetts, U.S.A.

Andrew Davis, M.D.
Director, Parasitic Diseases Programme, World Health Organization, Geneva, Switzerland

Carlos Gitler, PH.D.
E. Stanley Englund Professor of Membrane Biology, Department of Membrane Research, The Weizmann Institute of Sciences, Rehovot, Israel

William Goffman, PH.D.
Professor, School of Library Sciences, Case Western Reserve University, Cleveland, Ohio, U.S.A.

Chamlong Harinasuta, M.D., D.SC., D.T.M. PH.D.
Dean, Faculty of Tropical Medicine, Mahidol University, and Coordinator, SEAMEO-Tropical Medicine Project, Bangkok, Thailand

Adetokunbo O. Lucas, M.D.
Director, Special Programme for Research and Training in Tropical Diseases, World Health Organization, Geneva, Switzerland

Mao Shou-Pai, M.D.
Professor and Director, Institute of Parasitic Diseases (Shanghai), Chinese Academy of Medical Sciences, People's Republic of China

Adolfo Martínez-Palomo, M.D., D.SC.
Head, Section of Experimental Pathology, Centro de Investigacion y de Estudios Avanzados del Instituto Politechnico Nacional, Mexico City, Mexico

Graham F. Mitchell, PH.D.
Head, Immunoparasitology Unit, The Walter and Eliza Hall Institute of Medical Research, Melbourne, Australia

G. S. Nelson, M.D., D.SC.
Walter Myers Professor of Parasitology, Liverpool School of Tropical Medicine, Liverpool, England

Bridget M. Ogilvie, PH.D., D.SC.
Assistant Director, The Wellcome Trust, London, England

Fred R. Opperdoes, PH.D.
Head Research Unit for Tropical Diseases, International Institute of Cellular and Molecular Pathology, Brussels, Belgium

Peter Perlmann, PH.D.
Professor of Immunobiology, Department of Immunology, University of Stockholm, Stockholm, Sweden

M. D. Rickard, D.V.SC., PH.D.
Reader, Veterinary Parasitology, University of Melbourne, Weribee, Victoria, Australia

Michael Sela, PH.D.
Professor of Immunology, Department of Chemical Immunology, The Weizmann Institute of Sciences, Rehovot, Israel

Kenneth W. Sell, M.D., PH.D.
Scientific Director, Intramural Research Program, National Institute of Allergy and Infectious Diseases, Bethesda, Maryland, U.S.A.

Fiodor F. Soprunov, M.D.
Director, Martinsinovski Institute of Medical Parasitology and Tropical Medicine, Moscow, U.S.S.R.

Sir Kenneth L. Stuart, M.D.
Medical Adviser, Commonwealth Secretariat, London, England

Julia Walsh, M.D.
Assistant Professor, Division of Infectious Diseases, Department of Medicine, Channing Laboratory, Harvard Medical School, Boston, Massachusetts, U.S.A.

Kenneth S. Warren, M.D.
Director, Health Sciences, The Rockefeller Foundation; Professor of Medicine, New York University School of Medicine; Adjunct Professor, Rockefeller University; New York, New York, U.S.A.

Karl Western, M.D.
Assistant for International Affairs, National Institute of Allergy and Infectious Diseases, Bethesda, Maryland, U.S.A.

Peter Williams, M.B., F.R.C.P.
Director, The Wellcome Trust, London, England

Michael Worboys, D.PHIL.
Senior Lecturer in History, Department of History, Sheffield City Polytechnic, Sheffield, England

Contents

Preface

The science of parasitology is particularly relevant to the health of humans and animals throughout the world. As such, it provides the core of the Rockefeller Foundation program of biomedical research called The Great Neglected Diseases of Mankind. While some bacterial, viral, and fungal diseases are included in this program, the parasitic diseases caused by protozoa and helminths almost uniformly fit into this category. Many of them infect vast numbers of people and animals (of the order of hundreds of millions), but they are greatly neglected by the scientific establishment of the modern world. The principal reason for this neglect, both scientific and financial, is the fact that these diseases are primarily problems of the developing world. Hence, no vaccines exist for any of the human parasitic diseases; for many, treatment is either nonexistent or highly toxic; and diagnosis still relies largely on the visual examination of biopsies, blood, and excreta. In contrast, the discovery of a spate of antibiotics has revolutionized treatment of bacterial diseases, and vaccines have been developed for many bacterial and viral diseases. Furthermore, the field of microbiology, which deals largely with bacteriology and virology, has sparked the scientific revolutions of immunology and molecular biology.

In order to examine the present status of parasitology and the reasons for its putative neglect, as well as to develop means of remedying the situation, a series of conferences has been convened, bringing together classical parasitologists and a new and complementary breed of parasite-oriented immunologists, biochemists, molecular biologists, and population ecologists. Discussions focussed on history, teaching, research, the literature, training, career opportunities, funding, and the future of the field. Opportunities for a veritable renaissance in parasitology were highlighted by descriptions of new developments in cell biology, pharmacology, biochemistry, immunology, epidemiology, and ecology.

The first conference, entitled "The Current Status and Future of Parasitology," was jointly sponsored by the Josiah Macy, Jr., and The Rockefeller Foundations in New Orleans in October 1980. This meeting, which involved only scientists from the United States and Canada, resulted in a paperback publication available from the Macy Foundation.

In June 1981, a one-day meeting of donor agencies (both public and private) and scientists interested in the "new parasitology" was convened at The Rockefeller Foundation in New York. It was there that Jonas Salk coined the phrase, "the great neglected opportunities." This was the beginning of the remarkable new program in the Biology of Parasitism announced in July 1983 by the John D. and Catherine T. MacArthur Foundation.

The present publication is the result of a meeting held at The Rockefeller Foundation's Study Center in Bellagio, Italy, on April 22–29, 1982, to consider the status of parasitology globally. The participants were a blend of investigators working in the most advanced areas of modern science, leaders of global and regional programs in the developed world (U.S.A., Europe, U.S.S.R, Australia) and the developing world (China, Southeast Asia, Latin America, the Caribbean, and Africa), and representatives of major global funding agencies, both public and private. Papers were commissioned for presentation at the conference, and the ensuing discussions were completely recorded by stenotyping for publication of a full report as a benchmark in global parasitology.

We are grateful to the participants, who cooperated so well in the preparation of these papers and discussions.

1

The Emergence and Early Development of Parasitology

Michael Worboys

Introduction

Although I began my career as a biologist, I moved after my first degree to history and social studies of science. My main research interest has been in the history of science policy and scientific assistance to developing countries. When I began this work some ten years ago, the subject of "science for development" was attracting a lot of interest amongst sociologists and political scientists, yet all of the literature started with the great United Nations Conference in 1963. It was a subject without a history and I set out to write that history. I was aware when I started that much scientific work had been pursued in the "South" from the 19th century and earlier, but such was the quantity of material I found, that my original intention to deal with matters right up to 1963 had to be abandoned. I have since concentrated on the British Colonial Empire in the period up to 1940. It is events up to that date that I shall discuss in this paper, although with a wider geographical focus.

As well as being an honour, preparing this paper posed me with three problems. First, I move mainly amongst historians of science and medicine, so while I feel that I can gauge what subjects and problems will interest them, I can only hope that what I have to say will interest parasitologists, either "classical" or "new wave."

My second problem was that it proved difficult to place parasitology in the context of the development of the biomedical sciences in this century. The reason is simple: professional historians of science and medicine still ignore the 20th-century.[1] This in 1982, in the century when everyone agrees that science and medicine have made their most rapid progress and have had the most tremendous social impact. Moreover, what literature there is on medicine is typically "national" and ignores another obvious feature of 20th-century science and medicine, namely, its growing international dimension. A couple of months ago a colleague asked me how this paper was going; I replied that it was involving me partly in rewriting the whole history of the biomedical sciences in the 20th century. I was only half joking! Thus, much of this paper is speculative, the more so because I shall try to discuss parasitology in international terms. It is an attempt to sketch the broad contours of the emergence and early development of parasitology, not a definitive history. Having read the report of the recent American conference and spoken briefly to Dr. Warren some months ago, what I decided to do was to try to give historical answers to two questions that have interested parasitologists in the reassessment of their discipline over the last decade or so.[2]

First, the question, "What is parasitology?" Problems about the nature and boundaries of discipline are in fact as old as parasitology itself, as is the issue of whether it ought to be split up into protozoology, helminthology, and entomology. Thus, I shall discuss how and why a discipline called parasitology emerged at all.

The second question I want to consider is, why does it seem that parasitology has remained outside of the rapid progress of the biomedical sciences in this century? Thus, I want to assess the development of parasitology through to the 1930s and suggest reasons as to why it missed out of the microbiological revolution.

The two questions of course are linked, and I want to discuss them together by focusing on three periods in the development of parasitology. First, the period from the mid-19th century up to the turn of this century—what I shall call the "prehistory of parasitology." Here I want to say something briefly about the rise of scientific medicine and about helminthology, which was already a well established subject in the 19th century. Next, I want to look at the "emergence of parasitology" in the period 1900–1918. The influence of tropical medicine is an obvious feature of this period. However, I want to argue, against most other accounts of the discipline, that it did not fully establish itself as a separate and distinct discipline at this time, but rather that it was a focus of interest in still relatively undifferentiated biomedical sciences. In modern jargon, the biomedical sciences were interdisciplinary. Finally, I shall discuss the "establishment of parasitology" in the inter-war period. Here I shall chart how its "relative" decline and isolation went hand in hand with its institutionalization and the consolidation of its links with classical zoology, which had served it well before 1914, but which after 1918 isolated it from the new biology.

The Social History
of Scientific Disciplines

Before discussing these three periods I would like to say something briefly about my approach and about the study of scientific disciplines in general. The popular view of the history of science and medicine is that of a succession of heroic discoveries and discoverers: in science—Newton, Lavoisier, Darwin, Einstein; in medicine—Pasteur, Lister, Koch; in parasitology—Leeuwenhoek, Rudolphi, Steenstrup, Du Jardin, Leuckart, Laveran, Ross, Grassi. . . . On the whole, professional historians of science and medicine have never worked along these lines. What they have largely done is to study the development of scientific ideas and their evolution towards modern knowledge.[3] Such an approach is now out of favour for two reasons. First, it imposes upon the past the modern organization and interests of science, obscuring the fact that the past was different; that is, that science was pursued in different ways, with different objectives and in different conditions. Second, this approach assumes the autonomy of science and medicine; that is, that they developed by their own internal logic independent of other forces. It is now accepted, with obvious qualifications, that the further back in time one goes the less autonomous science and medicine were. Nowadays, historians are seeking to place scientific development in its historical and social context: in the context of the social system of science, and in the context of society at large. The new approach is often called the social history of science or medicine, in that it attempts to understand and relate scientific change to social changes.

One approach in this new social history is to focus on the emergence and development of scientific and medical disciplines.[4] After all, it is into a "discipline" that scientists are initiated as undergraduates, as researchers it is to a disciplinary body of knowledge that they contribute, and as professional scientists, while having a general allegiance to science, one's primary allegiance is to a discipline. While the study of disciplines is still in its infancy, a number of simple but important results are apparent. There are and have been a bewildering variety of disciplines. They have had various origins. They vary in size. They vary in type of subject matter. They may or may not change over time. They vary between countries, within a country, and even within the same university or research institute! Such diversity is all grist to the historian's mill. However, one general conclusion is clear: that disciplines are best studied as *social institutions*. We can ask questions about the social origins of disciplines as social institutions as well as about scientific origins. Subsequently, we can inquire about their social structure and about social relations and commitments. Emergence is, however, of crucial importance, because like any social institution, disciplines are conservative and patterns once established may prove difficult to shift. As one sociologist observed recently:

> [Once a discipline has emerged] vested interests are immediately gen-
> erated which tend to maintain it. Labouriously acquired competences
> and procedures are valued and not abandoned lightly: accepted laws
> and definitions become emblems of prestige and standing as well as
> technical resources.[5]

Of course, such conservatism or "dogma," as Kuhn calls it, is highly functional in the critical evaluation of new ideas.[6] At the same time science has
been and is characterized by rapid changes in ideas; so sociologically it is
very interesting, so much so that sociology of science is one of the fastest
growing areas in sociology. With regard to scientific change and innovation, what sociologists and historians have come to argue is that *mobility*
is perhaps the single most important factor.[7] By this they mean a number
of things: when new methods, techniques, and approaches are brought
into a discipline by migration; when scientists move to explore a new area
of ignorance; when changes are brought about by individuals "marginal"
to established subjects; when "external" social needs stimulate the exploration of a new or neglected topic.

How does all this help us understand the emergence of parasitology and
its subsequent development, or lack of it? It tells us that we must regard
our subject matter, parasitology, as problematic. We must not prejudge its
content or nature. It tells us that we must be aware of both scientific and
social factors in its emergence, that the pattern of emergence may be different in different locations. It tells us that once a discipline is established
we need to look at the forces that sustain it in a particular way. What follows is not, therefore, a history of the great discoveries in parasitology or
their impact, scientific or practical; rather it is a social history of a
discipline.

The Prehistory of Parasitology

In order to fully understand the emergence of parasitology in the 20th century one needs to know something about its prehistory. Potentially this is
a huge topic, but one well covered in books by Hoeppli and Foster, so I
shall confine myself making four points about the biomedical sciences in
the late 19th century.[8]

First, while the last quarter of the 19th century did see the advance of
specialization in biology and medicine—one historian talks about this
being a time of "scientific subdivisioning"—it seems to me that this tendency has been exaggerated by most historians.[9] I consider that in this
period the biomedical sciences constituted a common or unspecialized
scientific culture. It was easy to move from one subject to another, with
considerable academic and geographical mobility. The whole approach
was what we would now call interdisciplinary. One can illustrate this by

reference to almost any late 19th-century biomedical scientist, but I have chosen five so-called parasitologists. In Britain, the first great "parasitologist" was Spencer Cobbold.[10] He was a medical graduate who followed a career in, what, biology? Well, he held simultaneous appointments in botany and comparative anatomy, while researching helminths! He later became Professor of Botany and Helminthology, a seemingly curious combination. G. H. F. Nuttall, the founder of the journal *Parasitology*, was a medical graduate from California who again followed a research career.[11] He worked at Johns Hopkins and in Berlin, Baltimore, Gottingen, and Cambridge. His rather embarrassed obituary in the journal he founded noted that he had worked in serology, immunology, bacteriology, biochemistry, entomology, hygiene, tropical medicine, and parasitology.[12]

H. B. Ward was a natural science graduate who went on to do research in Germany, Italy, and France before returning to the United States to work for the Great Lakes Fisheries Commission.[13] Later at Nebraska he was Professor of Zoology and Dean of the Medical School—a zoologist! C. W. Stiles was another biology graduate who went off to Europe, where his researches included physics, physiology, zoology, anatomy, and bacteriology.[14] Back in the United States he settled into veterinary science, and then became American scientific attaché in Germany. He returned to the United States Public Health Service and went subsequently to work on Rockefeller's "Hookworm Program." In France, Laveran, another medical graduate this time, undertook a doctorate in anatomy before following a career in military medicine and administration, working only part-time on malaria. It is, I think, not only difficult but wrong to even try to locate any of these exclusively in a single disciplinary history.

The second point I wish to make is implicit in what I have just said, namely that biomedical "research" was being pursued widely at this time. The focus of activity and the model to be emulated was provided by universities and institutes in Germany and to a lesser extent by France—both Meccas for biomedical scientists. A research career was precarious, however, and the number of permanent full-time research posts was small. Thus, Robert Koch began his important research part time in an alcove in his consulting room or in the Department of Botany at the local university. Laveran, Manson, and Ross all pursued part-time research. The Cuban Carlos Finlay, having trained in France, Germany, and America, combined private medical practice and research. Foster in his book *The History of Parasitology* has argued that the great discoveries of parasitology were made by isolated individuals who were essentially "amateurs." He is wrong on two counts. The individuals may have been geographically isolated, but there were dense networks of communication facilitated by mobility and the absence of specialization. Second, the terms *amateur* and *professional* are historically inappropriate, biological and medical research being differently practised and constituted then.

What had stimulated research more than anything else was the growing acceptance of the germ theory of disease.[15] Through the 1880s and 1890s the biomedical community was increasingly unhappy about any disease whose cause remained unexplained. Moreover, there was a considerable reputation to be made in discovering a particular etiology, while the technical resources—the theory itself, the microscope, and simple laboratory procedures—were widely available. While the germ theory triumphed largely in scientific terms, however, in practical terms it was successful immediately only in surgery. The theory itself did not offer doctors any startling new therapies and the immediate results of bacteriological research and vaccine development were disappointing. Its impact on infectious diseases was felt to lie through preventive medicine and with success likely in the medium term, and then only if public agencies would take the necessary steps in sanitation and public health.

The final point I want to make in this section is that there is good evidence to suggest that the rapid acceptance of germ theory was due to the groundwork prepared for it by work on helminths in the mid-19th century. Farley's work on the spontaneous generation controversy and germ theory has shown that the acceptance of the "alternation of generation" and the conviction that pathogenic worms had specific origins rather than arose spontaneously did much to ease the birth pangs of germ theory.[16] In fact, the study of cestode and tremadote parasites was well advanced in the late 19th century under the title of the Entozoa or Helminthology, but not Parasitology. Its ultimate association with protozoology is in some way historically surprising. At the turn of the century the divisions between pathogenic viruses, bacteria, and protozoa were not clear cut. For example, sleeping sickness in man was first reported as of bacterial origin, while syphilis was reported as due to pathogenic protozoa. Yellow fever was for a long time thought to be a bacterial disease. The one certainty with it after 1900 was that it was insect-borne, a fact that allied it closely in the eyes of contemporaries to malaria. Two questions arise then: How did helminths get lumped together with protozoa in parasitology, or conversely, why were not pathogenic protozoa incorporated into a discipline wider in scope than what became bacteriology?

The Emergence of Parasitology

The key to the answer to both of these questions lies in the period 1900–1918 and what was undoubtedly the main impetus to the separate emergence of parasitology—the emergence of tropical medicine. Again I want to stress that we need to look at the scientific and social origins of this subject; indeed social factors appear to have been of primary importance. Equally, we need to recognize the nonemergence of the subject, in certain Latin American and Asian countries, where tropical diseases did become an integral part of medicine. Tropical medicine was, therefore, a subject that emerged in temperate, "Northern" countries. It emerged in these countries, not as part of the growing division of labour in medicine, but as an expedient when the newly discovered tropical diseases largely failed to gain a place in the medical curriculum. Again, there are differences. In France, which regarded its Empire as an integral part of the nation, the new diseases were taught in medical degrees. Politics, not science, determined the curriculum. Indeed, the French were much more honest when they began by calling the subject "colonial medicine."[17]

In the late 1890s and early 1900s the problem of disease in tropical colonies attracted enormous interest, even at the highest political level. It was in the words on one statesman, "a means of promoting imperial policy."[18] Why so much attention? For most Northern countries, neither tropical colonies nor tropical disease were new, both going back to at least the 17th century. In fact, right up until the 1890s high mortality rates amongst Europeans in the tropics were accepted as inevitable, although there was a large literature on tropical fevers and advice on how to avoid or combat them. What changed around 1900? First, growing international economic and political competition amongst the major capitalist countries led to an interest in colonies as potential new markets and sources of raw materials and emigration. The perceived need, therefore, was to develop/exploit colonial territories more systematically, largely by encouraging trade and investment. The high mortality rates amongst Europeans, however, and the popular image of the tropics as dark, hot, pathogenic, and generally hostile inhibited economic activity. Disease was seen to be the major factor holding back development/exploitation. The second change, however, was that economic interest coincided with the successes of germ theory generally, and specifically with the major diseases of the tropics: filariasis by Manson in 1879; malaria parasite by Laveran in 1880; *Vibrio cholera* by Koch in 1883; plague bacillus by Kitasato and Yersin in 1894; malarial life cycle by Grassi and Ross in 1897; trypanosomiasis by Bruce during 1896–1902 and by Chagas in 1908; leishmaniasis by various researchers during 1900–1911. The catch-all term *tropical fevers* disappeared, and with its demise seemed to come the possibility of conquering tropical diseases. Patrick Manson put the view of biomedical community in 1898 this way:

> I now firmly believe in the possibility of tropical colonisation by the white races. Heat and moisture are not in themselves the direct causes of any important tropical disease. The direct cause of 99% of these dieases are germs. . . . To kill them is simply a matter of knowledge and the application of this knowledge in sanitary science.[19]

The economic and political importance of this sort of statement was obvious, and the new discoveries were taken up, perhaps more rapidly than any other comparable scientific breakthrough in history. Indeed, the political aspect of the subject influenced the scientific community, for the new discoveries were accompanied by what one can only call "The Scramble for African Diseases" every bit as strong as the political rivalry—that is, priority disputes (some of which go on until this day), rival expeditions, competition for honours, and often acrimonious professional and personal relations.

The initial hope of those working with the new tropical diseases was that they would be integrated in existing medical institutions and teaching. In France, Latin America, and India this is what happened, but in Germany, Great Britain, and the United States, three of the four dominant scientific nations, integration failed and tropical medicine was institutionalized separately. The arguments for integration were powerful ones—the intrinsic importance of the subject, the growing number of doctors practising in the tropics, and the growing likelihood, with improvements in communications, of the importation of tropical diseases into the North. These arguments failed for two main reasons: The first was the intransigence of the medical establishment of curriculum reform, not least because by 1900 few of the major disease problems of the North were caused by pathogenic protozoa and helminths. Thus, while bacteriology developed in Northern hospitals, medical schools, universities, and research institutes, parasitology did not. Second, at this early stage tropical medicine had a strong entomological emphasis and was alien to the whole nature of orthodox medical education, a fact compounded by the very different approaches to disease control. With bacterial disease, researchers worked on vaccination and immunological studies, while with protozoal and helminthic diseases all of the emphasis was on vector control. It is important to remember just how large "the fly" loomed at this time (see Fig. 1-1)! There was, however, an urgent need to supplement medical training to equip Northern doctors for tropical or colonial practice. Up to 20% of British medical graduates were practising in warm climates. Thus, in an important sense tropical medicine was defined initially by what an orthodox medical degree left out. Quite simply this was: protozoology, entomology, tropical helminthology, basic microscopy, and tropical public health (netting, drainage, fly control, etc.).

Although the discipline was a practically defined one, attempts were made to justify it scientifically. The most interesting was Manson's attempt to distinguish bacterial or cosmopolitan diseases from parasitic or geographical disease,[20] the former being potentially universally distributed throughout the world, whereas the latter were dependént on the association of man and specific vectors and so of limited geographical distribution. Of course, such a division in no way mirrored tropical practice, where both classes of disease were prevalent. There was also the problem that many so-called tropical diseases—malaria, cholera, typhus, the plague— had often only recently been prevalent in Europe and North America. Another important factor in its emergence was that many university zoologists migrated to the study of pathogenic parasites. They did so because

FIG. 1-1. Mr. Austen Chamberlain, M.P. (Chairman of the Committee formed to raise 100,000 for the London School of Tropical Medicine Fund) loq.: "Come, John, you're wasting time and valuable lives. You must face facts. Either give us a little money, or your most promising sons will continue to die, and your trade will suffer—now, then, which shall it be?" (From Tropical Life 9 (1913): 32. This cartoon is reprinted by kind permission of the Bodleian Library, Oxford, England, as per reference 2323.d. 48 (9) 5 1913. p. 32.)

it was intellectually exciting—germ theory, the life cycle and the alternation of generation, and the microscope were powerful research tools. Added to this was the strong likelihood of identifying a new species, which guaranteed a publication, prestige, and perhaps immortality! Equally, it was a subject that was practically important, so it was one area in which zoologists could obtain funding. It must be remembered that at this time, in most countries, the biological sciences were small and poorly funded. So we find many university zoology departments building themselves up at this time with funds from tropical disease and veterinary agencies. It is important to stress that it was zoologists, not biologists, who migrated into parasitological work. In the early 20th century the division between botany and zoology was becoming well established. Conventionally, zoologists organized their work and teaching around "types," that is, representative species from each phylum from the protozoa upwards. Thus, the causative agents of the tropical diseases and their vectors were readily integrated into this framework. A consequence of the botany–zoology division was the fact that bacteria and viruses were regarded as plants—what Nuttall once called "vegetable parasites." Of course, such a distinction was merely conventional, but within universities it was a powerful one. One consequence of the new interest in tropical diseases was that scientific, medical, and veterinary journals were flooded with articles on parasitic protozoa, helminths, and insect vectors. Much of the work was in the tradition of classical zoology—taxonomy, morphology and life cycles—yet this was eminently practical, in that it facilitated the identification of specific pathogenic parasites and aided the attack on vectors by helping identify the vulnerable point in the life cycle. It could rightly claim to be applied zoology, or as the Americans termed it, "medical zoology."

I believe, however, that while something we can recognize as parasitology was emerging in this period, through its links with tropical medicine it remained very much part of the biomedical sciences. In this period there were virtually no exclusively parasitological institutions established. Rather, parasitology was practised in either schools or institutes of tropical medicine where the departmental organization was usually protozoology, helminthology, and medical entomology, or in veterinary institutions, or in university zoology departments. Moreover, I have come across little evidence of individuals calling themselves parasitologists at this juncture. There were few parsitological journals, the first English language journal, *Parasitology*, was founded in 1908 as a supplement to the *Journal of Hygiene*. Major articles were published in a variety of journals, for example Leiper's classic work on schistosomiosis was published in general medical, zoological, and military medical journals, not in journals of tropical medicine or parasitology. This was even more the case in Latin American and Asian countries, where separate tropical medical institutes failed to emerge. Indeed, in Brazil, as Nancy Stepan has pointed out, the perils of early specialization are reflected in the fact that a Bacteriological Institute established in Sao Paulo failed, whereas the Oswaldo Cruz Institut in Rio de Janeiro was successful because it maintained a wide range of pure and applied research and medical services across the board.[21]

More generally, I think it could be argued that the study of tropical diseases was not just another part of biomedicine, but was a leading part before 1914. Two of the first ten Nobel Prizes for Physiology and Medicine (Ross and Laveran) were for work on malaria, and when Robert Koch and Paul Ehrlich received theirs in 1905 and 1907, respectively, both were working on problems related to tropical diseases. There were the successes of vector control with yellow fever, individual malarial precautions, and the attempts at mosquito eradication. Parasitic diseases were at the forefront of chemotherapy. To quote Singer and Underwood:

> [Until the mid 1920s] the only chemotherapeutic substances which were effective were mercury, quinine and its alkaloids [i.e., malaria], emetine [tropical dysentery], salvarsan [developed by Ehrlich when working on trypanosomes], and Bayer 205 [trypanosomes].[22]

Indeed, one of the first synthetic drugs was chloroquin, whose development was begun in Germany when quinine supplies from the Far East were threatened during the First World War. By comparison, even the much vaunted discipline of bacteriology may appear backward; its period of rapid development and direct success did not really begin until the 1930s. So perhaps one problem was that parasitology peaked too early.

The Establishment of Parasitology

What happened then in the inter-war period when other biomedical disciplines took off? Were tropical medicine and parasitology left behind and, if so, why? The period 1914–1940 is when I believe parasitology was "established"; that is, when scientists began to call themselves parasitologists, when parasitological institutes and associations were founded, when a parasitological education, usually postgraduate, became available, and when journals began to proliferate and were founded in the major scientific nations.

What I have to say in this part of my paper is even more speculative than what has gone before, because in researching this period it became clear that few historians of science and medicine have studied this crucial time in biomedical advance. There are heroic biographies of Domagk, Fleming, Florey, and Chain, one or two new studies of molecular biology, of the "Road to DNA" type, and some studies of the work of the Rockefeller Foundation in promoting scientific medicine.[23] There is, however, little that is related to the medical aspects of microbiology and its impact, so it is somewhat difficult to place parasitology in context. Moreover, it is in this period that parasitology became truly "international," as part of a seeming convergence in the patterns of scientific activity worldwide. Historians and sociologists of science have ignored this feature. Indeed, the emphasis in current work on *local* traditions and national differences leaves the "internationale" of science as a real problem. We don't know whether convergence was real or, if it was real, whether it developed "naturally" out of scientific communication or whether certain countries were scientifically dominant and their patterns copied, or whatever.

What we do know is that after 1918 political and biomedical interest in tropical medicine declined, to revive again in the Second World War, but at no time since has it enjoyed the importance and prestige it had between 1900 and 1914. This relative decline in its standing was caused by many factors. Politically the First World War and its aftermath—revolution, depression, recession, and rearmament—meant that the main issues of the day were exclusively Northern. The problems of the South and of the tropical colonies were given a lower priority by the major Northern powers. The problem of European health in the tropics was felt to have been more or less solved and became routine. In the colonial territories attention shifted to the health of the indigenous population, but any improvement here was seen to have to be long term, especially as often expensive vector-control programmes were the main approach. Indeed, it was not until the 1920s and 1930s that any detailed picture of the health position of the population of the South was established.[24] What emerged was a picture of endemic disease and malnutrition, which seems to have surprised many people at the time. This led at one level to a reevaluation of the origins of ill-health in the South, with the focus shifting from the agents of disease to nutrition and environment. At a meeting of representatives from medical services from Africa and India in 1932, the following typical resolution was passed:

> In an undernourished population . . . the mere treatment of disease, no matter how effective and widely carried out, will achieve negligible results. The first need is a continuous supply of sufficient and well balanced food for the nature to resist infection, and the next is the improvement of housing.[25]

Many doctors in colonial territories felt they were dealing with a population whose general health was declining because of malnutrition. The anthroplogist Audrey Richards observed in 1939 that the "African diet had deteriorated in contact with white civilization rather than the reverse."[26] These views are interesting because many historians of medicine in Northern countries believe that the long-term improvements in health and longevity over the last century or so have been due largely to improvements in diet and the environment.[27] The assumption is that too much of the credit has been claimed by modern medicine and that there has been a confusion of scientific with medical success.

Another change, associated with the relative decline in the standing of tropical medicine, was its decentralization within the major European Empires. The build-up of medical services in the colonies reached the critical mass whereby they needed and could support their own research laboratories. At the periphery, however, research seems to have become largely "technical service," and longer term research was neglected. The schools and institutes in the North appear to have largely marked time, although much important work was done. In an large sense, researchers in tropical medicine became more isolated—the whole discipline moved from a heavy research orientation to routine, the health problems of the tropics lost their political urgency, researchers were dispersed rather than clustered, and they began to lose touch with other medical and biological subjects.

The period after the First World War in most countries saw rapidly increasing funding for science and medicine, as shown in the growth of universities, the proliferation of research institutes, the growth of industrial R and D, and government departments for research. In medicine, it has been suggested by Rene Dubos, the pandemic of influenza in 1918–1919—which killed 20 million people worldwide, more than died in the war itself—did most to galvanize action in Northern countries.[28] Government funds for medical research were controlled on the whole by the medical establishment in each country and were channelled in clinical research and postgraduate medical schools—in short to Northern problems. Thus, bacteriology entrenched in medical schools and hospitals was a beneficiary. Such growth in itself, together with the growth of applied research, facilitated specialization. My belief, then, is that specialization occurred in this period, somewhat later than is generally stated. There was little new Northern government money for tropical medicine or parasitological research, and support was left to private philanthropy. Of course, the Rockefeller Foundation played a crucial role at this time through the work of the International Health Board. While the Rockefeller Foundation was supporting work on malaria and hookworm, however, it was generally out to promote the discipline of "hygiene" and education in public health and disease prevention. In Britain, although Leiper managed to associate the London School of Tropical Medicine with the planned Institute of Hygiene and got the present school built in the 1920s, through the 1930s there was a lack of funding.[29] Similarly at Cambridge, Nuttall got £30,000 from Sir Percy Molteno, an Indian tea merchant, for the Institute that now bears his name, but he could not obtain funds for staff or research.[30] Indeed after Nuttall retired, the Institute became, under David Keilin, famous for biochemistry and work on the cytochrome—so much had "applied"/"practical" pressures declined.[31]

Along with increased research funding went the growth of higher education and postgraduate research—still small by today's standards but nonetheless enough to consolidate specialization. After 1918 the historical links between botany and medicine, via the *Materia Medica*, and zoology and medicine, via comparative anatomy, physiology, and even parasitology, were broken. Medical schools and faculties became autonomous, and zoology and botany became university subjects in their own right. Ironically, at exactly the time of this serverance, new links between medicine and biology were being reforged, only by chemists and physicists moving in to revolutionize the biological sciences through biochemistry, cell biology, and microbiology.

Scientifically, by 1918 the morphology and life cycles of most of the important tropical parasitic diseases were known; there were few prestigious discoveries left to be made. The problems were now more difficult, being related to the control of disease, pathogenicity, and epidemiology. In the laboratories, parasitologists did not have the advantage of working with such cooperative organisms as bacteria. Certainly the cutting edge was very much in vector control, a trend that did much to advance the growth of entomology as an autonomous discipline, although other arthropods (ticks and mites) remained within parasitology. Overall, parasitology was now staffed by second generation scientists, who had not come to the subject via the late 19th century interdisciplinary route; concentration on a problem in one generation of scientists produces specialists in the next.

In university zoology departments, the tools of classical parasitology remained as powerful as ever and zoologists turned them on any and every parasite they could find. The prestige of discovering a new species or the detective work in elucidating life cycles provided exciting, readily publishable, and seemingly endless opportunities for research. The study of parasites became an end in itself, rather than a means to an end as it had been before 1914. By the 1930s only about 20% of articles in, say, *Parasitology* were on important medical or veterinary species. Without the impetus of applied needs, its classical approach, still extremely fruitful in its own terms, became frozen.

All of this is complicated by national differences. In Europe, parasitology's link with tropical medicine remained important. In the United States the links with veterinary science and public health were as important as, if not more important than, those with tropical medicine.[32] In Latin America, Asia, and even India, otherwise subject to scientific imperialism, it seems that parasitology remained part of the general biomedical sciences. Perhaps the relatively small size of these scientific communities and the fact that tropical diseases were not exotic were important here. Thus, it would seem that parasitology established itself as a largely Northern discipline. By the end of the 1930s there were parasitological journals in all of the main Northern countries, but the proliferation of journals worldwide does not seem to have come until after 1945. Parasitology did become established, however, and it achieved this as much from weakness as from strength.

I would like to end this section with what I must admit are two guesses about the establishment of parasitology worldwide at this time. First, I feel that parasitology which, as the combination of protozoology, helminthology, and entomology, had been defined by tropical medicine and consolidated by zoological connections, was perpetuated because of the still small size and weakness of the subjects. Only in consultation could a viable educational curriculum and research network be sustained. In this period what few parasitological journals there were did not expand greatly in either size or circulation. Second, I feel that international convergence in the content and boundaries of the discipline was due to the scientific dominance of the major Northern countries, in terms of channels of communication, educational opportunities, funding, and the model they provided emergent scientific communities—in short, to scientific imperialism.

Conclusion

At the beginning of this paper I set out to give historical answers to two questions: What is parasitology? and Why does it appear to have been left behind by other biomedical sciences since the 1930s? In answer to the first question, I have suggested that parasitology emerged to provide the zoological underpinning for tropical medicine. Its subject matter was defined on the one hand by the tropical medicine curriculum and on the other by the botany–zoology division and the way tropical parasites fitted neatly into the "type"-based zoological curriculum. Moreover, in the early 20th century, classical zoological work on taxonomy and life cycles was eminently practical. I stressed that in the period of its emergence, parasitology can be seen to have been at the forefront of a still highly mobile and largely interdisciplinary biomedical scientific community. The relative decline in the importance and standing of parasitology ironically came with its institutionalization in the inter-war period. Tropical disease problems lost their urgency, tropical medicine became routine, and research concentrated on vector control. The general postwar growth in science and medicine passed by parasitology. The research programme of taxonomy, morphology and life-cycle elucidation, however, remained viable and gave zoologists powerful resources for classical work. The absence of external pressures allowed this fruitful research programme to perpetuate itself. It remained a field with applied medical and veterinary connections, but generally the study of parasites had become an end in itself. Elsewhere, the major developments in the biomedical sciences were coming *and* still to come from biochemistry, chemistry, and physics, not zoology.

Nonetheless, it would be wrong to underestimate the important advances made in tropical medicine and parasitology in this period, with often very limited resources. I am talking of "relative decline," so that parasitology did continue to make progress, but not as rapidly as other biomedical disciplines. The contribution of classical parasitology to disease control is difficult to judge, not least because research was not always applied consistently and effectively. Its impact, however, could be enormous.[33]

In order to make these conclusions about the emergence and early development of parasitology I have had to discuss, I am aware, very schematically and unevenly the development of five major subjects—parasitology, tropical medicine, medical research generally, zoology, and veterinary science—over a period of 50 years, embracing many countries. Clearly my subject warrants a book or series of volumes rather than a single paper. Thus, much of what I have said must be incomplete and tentative. The range of subjects and issues I have had to discuss, however, reveals that the development of the history of parasitology has as much to offer the history of science and medicine in developing new, interdisciplinary approaches as parasitology has to offer current scientific and medical research.

References

1. Parasitology is, in fact, better served than many other biomedical disciplines with regard to its history in the 20th century. Many valuable studies have been produced on particular discoveries and individuals by senior or retired parasitologists. See: Scott HH (1939) A history of tropical medicine. Edward Arnold, London; Foster WD (1965) A history of parasitology. E and S Livingstone, London; Garnham PCC (1970) Progress in parasitology. Athlone Press, London; Chernin E (ed) (1977) Milestones in the history of tropical medicine and hygiene. Am J Tropical Med Hygiene 26: 1053–1104; Moore DV (1976) Fifty years of American parasitology. J Parasitol 62: 498–514

2. On the reassessment of parasitology see: Ulmer MJ (1978) What's past is prologue. J Parasitol 64: 3–13; Taylor AER, Muller R (eds) (1978) The relevance of parasitology to human welfare today. Blackwell, Oxford; Warren KS, Purcell EF (eds) (1981) The current status and future of parasitology. Josiah Macy Jr Foundation, New York

3. The literature of the recent changes in the history of science and medicine is now enormous. Two recent reviews, both with comprehensive bibliographies, can be recommended as introductions. MacLeod RM (1977) Changing perspectives in the history of science. In: Rosling-Spiegel, I, de Solla Price, DJ (ed) Science technology and society. Sage, London; Woodward J, Richards D (1976) Towards a social history of medicine. In: Woodward J, Richards D (eds) Health care and popular medicine in nineteenth century England. Croom Helm, London, pp 15–55

4. On new disciplines see: Lemaine G, MacLeod R, Mulkay M, Weingart P (eds) (1976) Perspectives on the emergence of scientific disciplines. Mouton, The

Hague; Edge DO, Mulkay MJ, (1976) Astronomy transformed. Wiley, New York

5. Barnes B (1982) TS Kuhn and social science. Macmillan, London; p. 110
6. Kuhn TS (1963) The function of dogma in scientific research. In: Crombie AC (ed) Scientific change. Heinemann, London; pp 347–369
7. On the importance of mobility in scientific change, see: Mulkay M (1972) The social process of scientific innovation. Macmillan, London
8. Hoeppli R (1959) Parasites and parasitic infections in early medicine and science. University of Malaya Press, Singapore; Foster WD (1965) A history of parasitology. E and S Livingstone, London
9. Singer C, Underwood EA (1962) A short history of medicine. Clarendon, Oxford; Section VII, especially pp 205–208
10. See: Foster WD (1961) Thomas Spencer Cobbold and British parasitology. Med History 5: 341–348
11. See: George Henry Falkiner Nuttall, 1862–1937 (1938) Parasitology, 30: 403–415
12. Singer C, Underwood EA (1962) A short history of medicine. Clarendon, Oxford. Nuttall appears as an immunologist
13. See: Cort WW (1946) Obituary: Henry Baldwin Ward. Science 102
14. See: Cram EB (1956) Stepping stones in the history of the American Society of Parasitology. J Parasitol 42: 461–473; Cassedy JH (1971) "The germ of laziness" in the South 1900–1914: Charles Wardell Stiles and the progressive paradox. Bull History Med 45: 159–169
15. Crellin JK (1968) The dawn of the germ theory. In: Poynter FNL (ed) Medicine and science in the 1860s. Wellcome Institute for the History of Medicine, London; pp 131–150; Foster WD (1970) A history of medical bacteriology and immunology. Heinemann, London
16. Farley J (1972) The spontaneous generation controversy (1700–1860): The origin of parasitic worms. J History Biol 5: 95–125
17. For more detailed discussion of the development of parasitology in France, the Soviet Union, and China, see the chapters by Capron, Soprunov, and Mao in this volume.
18. Joseph Chamberlain, British Secretary of State for the Colonies 1895–1903. Speech made on 10 May 1899. Quoted in Worboys M (1980) Science and British colonial imperialism 1895–1940. PhD dissertation, University of Sussex, p 94
19. Manson P (1898) Discussion on the possibility of acclimatisation of Europeans in tropical countries. Brit Med J 1168. On the emergence of tropical medicine see: Worboys M (1980) Science and British colonial imperialism 1895–1940. PhD dissertation, University of Sussex, Chap 3; Worboys M (1976) The emergence of tropical medicine. In: Lemaine et al (eds) Perspectives on the emergence of scientific disciplines. Mouton, The Hague, pp. 75–98
20. Manson P (1899) The need for special training in tropical diseases. J Tropical Med 2: 57–62
21. Stepan N (1976) Beginnings of Brazilian science: Oswaldo Cruz, medical research and policy, 1890–1920. Science History Publications, London and New York
22. Singer C, Underwood EA (1962) A short history of medicine. Clarendon, Oxford, pp 687–697
23. On biomedical science see: Crellin JK (1968) The dawn of the germ theory.

In: Poynter FNL (ed) Medicine and science in the 1860s. Wellcome Institute for the History of Medicine, London, pp 131–150; Foster WD (1970) A history of medical bacteriology and immunology. Heinemann, London; on molecular biology see: Judson HF (1980) Reflections on the histiography of molecular biology. Minerva 18 369–421; on Rockefeller see Kohler RE (1976) The management of science: The experience of Warren Weaver and the Rockefeller Foundation Program in molecular biology. Minerva 14: 279–306

24. Worboys M (1980) Science and British colonial imperialism 1895–1940. PhD dissertation, University of Sussex, Chap 8

25. International Conference of Representatives of Health Services of African Territories and British India (1933). Q Bull Health Organization League of Nations 2 115

26. Richards AJ (1939) Land labour and diet in North Rhodesia—An economic study of the Bemba tribe. Oxford University Press, London, p 1

27. McKeown T (1965) Medicine in modern society. Allen and Unwin, London; Powles J (1973) On the limitations of modern medicine. Sci med man 1: 1–30

28. Dubos R (1959) Mirage of health. Harper, New York, p 70

29. Fischer D (1978) Rockefeller philanthropy and the British Empire: The creation of the London School of Hygiene and Tropical Medicine. History Ed 7: 129–143 Unfortunately Fischer does not explain how an institution for "hygiene" came to incorporate a school of tropical medicine, nor of the relative importance of the two sides with the new school.

30. Nuttall GHK (1922) The molteno Institute for Research in Parasitology, University of Cambridge, with an account of how it came to be founded. Parasitology, 14: 97–126; Nuttall GHF (1931) Concerning the Molteno Institute for Research in Parasitology, University of Cambridge. Privately printed at the University Press

31. Keilin D (1966) The history of cell respiration and cytochromes. Cambridge University Press, Cambridge

32. Schwabe CW (1981) A brief history of American parasitology: The veterinary connection between medicine and zoology. In: Warren KS, Purcell EF (eds) The current status and future of parasitology. Josiah Macy Jr Foundation, New York, pp 21–43

33. See Worboys M (1980) Science and British colonial imperialism 1895–1940. PhD dissertation, University of Sussex, chap 7, especially sections on trypanomiasis control in West Africa.

Current Status of Parasitic Diseases

2

Parasitic Diseases in the South (Developing World)

Chamlong Harinasuta

The Southeast Asian countries are situated in the tropical zone with a total population of about 340 million. They include Burma, Thailand, Vietnam, Kampuchea (Cambodia), Malaysia, Singapore, the Philippines, and Indonesia. These are considered as developing countries, with the rural communities comprising about 80% of the population. In these rural areas, there are many important endemic parasitic diseases affecting the health and well-being of the people. They include:

A. Protozoal Diseases
    1. Malaria
    2. Intestinal amoebiasis and amoebic liver abscess
    3. Giardiasis
B. Helminthic Diseases
    Nematode Diseases
    4. Hookworm
    5. Ascariasis
    6. Trichuriasis
    7. Strongyloidiasis
    8. Capillariasis
    9. Filariasis—malayan and bancroftian
    10. Angiostrongyliasis
    11. Trichinosis
    12. Gnathostomiasis
    Trematode Diseases
    13. Schistosomiasis—*S. japonicum, S. mekongi*
    14. Paragonimiasis
    15. Opisthorchiasis

16. Intestinal fluke infections—fasciolopsiasis, echinostomiasis
Cestode Disease
17. Taeniasis

The epidemiology of these 17 parasitic diseases with a brief account of
their preventive and control measures is described:

Malaria

Malaria is a major public health problem in developing countries of Southeast Asia, causing considerable morbidity and mortality especially in those
living in the rural areas near semi-forested and hilly regions where cultivation of rice and other agricultural crops has been expanded. Malaria as
a leading disease in all tropical countries in Southeast Asia causes a great
deal of economic loss, not only from death but also from the cost of treatment, prevention, and control as well as of man-days of work resulting in
reduction in agricultural products and other agro-industrial productive
activities. This is a very important issue to be considered since 80–85% of
the people in our developing countries earn their living by agriculture. It
is noted that even though every country in Southeast Asia has a malaria
control programme with effective results, malaria is still prevalent in some
particular rural areas, especially during the last 5–6 years, due to various
causes.

With the introduction of the use of DDT residual house spraying to kill
Anopheles mosquito vectors of malaria, and the detection and treatment of
malaria cases in 1950–1970, the mortality and morbidity of malaria in each
developing country had been reduced to a very low level, leading to satisfactory results with cessation of malaria transmission in the rural areas.
Since 1971, however, the incidence of malaria in Southeast Asia has grad-
ually increased, with an alarming rate during 1976–1980. This is due to
many factors including technical and academic, social and economic, and
administrative problems in the national malaria control programmes of the
respective countries. Those problems are being identified, studied, and
solved in order to fulfill the objectives of the national antimalarial
programmes.

The brief account of malaria status in each developing country is as
follows:

Burma

Malaria is recognized as the most important public health problem in
Burma. At present malaria ranks as a principal cause of deaths in hospitals
in Burma, while its morbidity in some rural areas is estimated to be 45 per
1,000 people. There are more cases of *Plasmodium falciparum* than *P.
vivax*, the ratio being 3:1 in 1979 and many cases are resistant to chloro-

quine. *Anopheles minimus minimus* is the main vector mosquito, while *A. annularis, A. sundaicus, A. culicifacies*, and *A. balabacensis* are principal vectors. It was found recently that *A. annularis* and *A. culicifacies* have developed resistance to DDT in some rural areas.

The launching of the Malaria Eradication Programme in Burma in 1960 gave encouraging results with a marked reduction of malaria cases and death rate. During 1970–1975, however, the malaria status in Burma deteriorated with a higher incidence of malaria cases. The infection was particularly prevalent in hilly forested areas where *A. minimus* and *A. balabacensis* predominated; in coastal areas with *A. sundaicus*; and in irrigated agricultural areas where *A. culicifacies* breed in abundance. The situation, however, has been improved since 1977 due to integrated control measures including DDT house spraying, case-finding and treatment, chemoprophylaxis, and bio-environmental measures.

In 1979 the slide positive rate (SPR) of malaria in Burma was 1.2%; out of 34.0 million population, 7% were living in malaria-free areas, while 93% were receiving protection against malaria by drugs, DDT house spraying, surveillance operation, and/or vigilance activities.

Thailand

Malaria is still considered to be a major public health problem in Thailand. It causes significant morbidity and some mortality in many population groups, especially those in the rural areas near and within the semi-forested and foot-hill areas along the borders with the neighboring countries of Kampuchea, Laos, Burma, and Malaysia, estimated to be about 9.3 million people in 1980. The main factor is a large-scale population movement (such as land settlements in new areas or seasonal agricultural labourer movement) precipitated by rural high population growth and social and economic pressure together with low agricultural productivity in northeast Thailand.

In 1980, the slide positive rate (SPR) of malaria in Thailand was 8.1 per 1,000 population (395,442 positive slides) with a mortality rate of 8.0 per 100,000 people (in 1947 the mortality rate of malaria was 286 per 100,000 population for the whole country).

There are more cases of *P. falciparum* than *P. vivax*, the ratio being 2:1 in 1980, and more than 90% of falciparum infections show resistance to chloroquine and to Fansidar. The principal vectors of malaria in Thailand are *A. balabacensis balabacensis, A. minimus minimus*, and *A. maculatus*, while *A. sundaicus* and *A. aconitus* are secondary vectors. *A. campestris, A. philippinensis*, and *A. culicifacies* are suspected vectors. All of these mosquitoes are still susceptible to DDT.

The Thai National Control Programme of Malaria opened in 1949, leading to a reduction of the mortality rate from 201.5 per 100,000 population in 1949 to 30.2 in 1969. After the operation of the Thai National Malaria Eradication Programme beginning in 1965, the malaria mortality rate was further reduced to be 10.1 per 100,000 population in 1970. In 1971–1974, however, the malaria situation in Thailand deteriorated, increasing the mortality rate to 12.5 and 15.8 in 1971 and 1974, respectively. The reasons were technical, social-economic, and administrative problems in the National Malaria Eradication Programme. Those problems have been identified and partially solved, resulting in some reduction in mortality rate, to 9.7 and 8.2 per 100,000 population in 1978 and 1979, respectively. The Eradication Programme has changed to a Malaria and Vector Control Project since 1975.

The antimalaria operation in Thailand consisting of DDT house-spraying in combination with case detection (active and passive) and treatment has resulted in a remarkable change in the epidemiological pattern of malaria in the country. In the flat flooded rice field and agricultural areas in central and south Thailand and other large areas of low or moderate malaria endemicity in the central part of northeast Thailand and the lower part of north Thailand, malaria transmission has been interrupted, resulting in a very low number of malaria cases. The positive cases are usually found to be imported from elsewhere. In the semi-forested and hilly areas near the borders in west, north, east, and south Thailand, however, there is still high and moderate malaria endemicity and *A. minimus* and *A. balabacensis* as the potent mosquito vectors. In these areas the antimalarial campaign has achieved partially successful results. The annual parasite incidences (API) per 1,000 population was 7.1–7.7 during 1975–1979 and 8.9 in 1980. The increase in the parasitic incidences is due mainly to the internal population movement, as mentioned above. The migrants to new areas usually build new huts and isolated dwellings in remote areas previously uninhabited. Residual house spraying therefore cannot cover all of them. Often the migration of these people has preceded the initiation of the antimalarial measures and thus a high rate of malaria infection is found among this group of the population.

Other factors in connection with the maintenance of still active transmission of malaria in some particular areas include (1) certain occupational habits such as rubber tapping very early in the morning in south Thailand resulting in more exposure to the biting of mosquito vectors, (2) man-made artificial breeding places for *A. balabacensis* created by gem mining in certain provinces in east Thailand, (3) increase in chloroquine resistance and sulfonamide resistance in falciparum malaria, distributed throughout the country (more than 90%), (4) changing habits of *A. minimus* and *A. balabacensis* mosquitoes to outdoor rather than indoor biting and resting, (5) less community participation in the DDT house spraying, complete spraying being made in less than 50% of the total households, and (6) an inefficient administrative sector due to insufficient funding of the antimalarial programme and related problems.

The Thai Government has tried to solve these problems, using technical and administrative approaches together with an increase in the budget for the antimalarial programme both by a loan and grant agreement with USAID and by national funds. In 1979–1980, a total of 42.7 million people were under the National Malaria Control Programme. This includes DDT house spraying, with other *Anopheles* control measures covering a population at risk of 9.3 million residing near or in semi-forested hilly border areas. Case detection and treatment in consolidation areas covered 2.2 million people, and surveillance integrated with provincial health services covered 31.2 million population. Bangkok, the capital city with 5.0 million people, is free from malaria.

Recently the Thai Government has established "Malaria Clinics" with microscopists to provide the rapid services of case detection and treatment, in areas of high incidence at the Region, Zone, and Sector Offices, and at some provincial hospitals and health centres. This has proved to be a very effective service and well accepted by the local people and communities.

Vietnam

There has been lack of information on the malaria situation in the Indochina countries of Vietnam, Laos, and Kampuchea (Cambodia), after marked political changes in their countries in 1975.

Vietnam has a total population of 53 million people in both the North and South sectors. Malaria is one of the main public health problems with a prevalence of more than 2.0 million cases annually. In Hanoi in the North, Ho Chi Min City (Saigon) in the South, and some other urban cities, there is no malaria. In other parts of the country, however, especially the rural and remote areas where 80% of the total population live, malaria is found to be prevalent. The hyperendemic areas of 10–50% are located in the semi-forested and forested mountainous jungle central highlands, while the meso-endemic and hypo-endemic areas of 3–10% are found scattered throughout the country. The coastal areas extending from the north to the south of the country have low malarial parasite rates. *P. falciparum* has been found to be the parasite in 50–60% of the cases and with marked resistance to chloroquine. The principal mosquito vectors are *A. minimus* and *A. balabacensis. A. jeyporiensis* is a vector found in the hilly and mountainous jungle covered areas, while *A. sundaicus* is the vector in coastal plains and in the Mekong Delta in the south. *A. maculatus* is suspected as another vector.

The national antimalarial programme created in 1953 has not given satisfactory results, because its operation was interrupted nearly every year by the war conditions and insecurity in the country especially during 1960–1975. Again between 1976–1980 the politicial condition of the country deteriorated, resulting in an ineffective antimalaria campaign in Vietnam.

According to WHO reports, the positive slide rate of malaria in Vietnam was 5.1% in 1979 as compared with 15–20% in 1975. At present the WHO

Special Programme for Research and Training in Tropical Diseases is assisting Vietnam in research and training on malaria in selected rural areas around Hanoi.

Kampuchea (Cambodia)

In 1982, Kampuchea continues to suffer unsettled conditions due to political dispute and war. Malaria is considered as the major public health problem, with its high prevalence in the mountainous forested parts of the country, and areas along the borders with neighbouring countries—Thailand in the north and Vietnam in the east. At camps in Thailand near the Thai–Kampuchea border with many hundred thousand refugees, malaria is found to be the major disease and most cases resist chloroquine and Fansidar.

According to WHO Reports in the past, malaria occurred throughout the year with positive rates of 20–29% in many areas; its peak incidences were found in the rainy season, June–September. The slide positive rates in some high endemic areas during 1969–1972 ranged from 27.3 to 31.2%. It has been estimated that about 80% of the country contains malarious areas. There was no malaria in Phnom Penh, the capital city, and in urban areas of some other towns in the central plains. Falciparum malaria was found to have a frequency of 45%.

The principal mosquito vectors are *A. minimus* and *A. balabacensis*, while *A. maculatus* and *A. sundaicus* are also important.

Although the national antimalarial campaign commenced in 1969 with DDT house spraying and drug treatment, the operation was interrupted from time to time by the political instability, military activities, and insecurity in the country. There is still a lack of information on the malaria situation at the present time in Kampuchea.

Laos

Laos has a population of 3.4 million people, 85% residing in areas at risk of malaria, since most of the country is hilly, mountainous, semi-forested, and forested areas. The Vientiane plain where the capital city of Ventiane is situated is considered a malaria-free area. Malaria has been found to be endemic throughout the country with varying incidences depending upon the prevalence of *A. minimus* and *A. balabacensis*, the principal vectors. Falciparum malaria has an incidence of approximately 53% and is resistant to chloroquine.

The antimalarial programme in Laos commenced in 1966 with DDT house spraying and case detection with drug treatment. The programme was interrupted from time to time, however, due to political troubles and war, insecurity, and lack of drugs, resulting in an unsuccessful campaign.

At present, little is known on the current malaria situation in Laos. However, WHO has reported that DDT house spraying was resumed in 1977 and extended into Vietiane Province in 1978. This report predicted further expansion of the program geographically in 1979–1980.

Malaysia

Malaysia consists of two regions: West Malaysia on the Southeast Asian Peninsula and the states of Sarawak and Sabah, to the east, on the island of Borneo. The total population numbers 13 million. In West Malaysia malaria has been recognized as a disease of public health importance in the last three decades due to the opening up of new lands for the cultivation of additional agricultural products. In 1965 the overall slide positive rate of malaria was 3.9%, while in some villages the infection rates rose to 21–90%, which could hamper their socioeconomic development. Thus the national malaria eradication programme was introduced in 1967, continuing until 1982. The undertaking was carried out in stages, starting from the north of Malaysia extending to the south covering eight regions of the country year by year. At present the whole of Peninsular Malaysia is under the control programme, resulting in the reduction of malaria cases from 400,000 per year during 1965–1967 to less than 11,000 cases in 1978. Falciparum malaria was found to infect about 65–76%. Many of those cases were resistant to chloroquine.

In East Malaysia, Sarawak with the population of one million has a malaria infection rate of about 1%, while Sabah with the population of 0.4 million has no problem in the control of malaria.

The principal mosquito vectors in Malaysia are *A. maculatus, A. sundaicus,* and *A. campestris,* while *A. balabacensis* and *A. letifer* are other vector mosquitoes.

Philippines

Malaria continues as a public health problem in the Philippines with 105,000 cases and 88,000 cases in 1978 and 1979, respectively, from a total population of about 45 million. The hyperendemic malaria areas are in the foothills, semi-forested, and mountainous areas, located on many islands in the southern part of the country. Active malaria transmission occurs mainly in the fringes of the forests where settlers are moving into newly opened agricultural lands for cultivation of crops. These mobile people usually live in temporary huts or dwellings with or without walls, which cannot be covered by the antimalaria campaign of the country. This condition is found on many islands in the Philippines, and the problem is very difficult to solve.

The national antimalaria programme commenced in 1955 when there were about 1.5–2 million malaria cases a year. Twenty-five years later the mortality and morbidity of malaria were reduced a great deal, even though in some areas as mentioned above incidences remained moderately endemic. The malaria infection consists of *P. falciparum,* 67%; *P. vivax,* 30.6%; *P. malariae,* 0.1%; and mixed infection, 2.3%. The chloroquine-resistant cases are distributed throughout the country.

The principal mosquito vector of malaria in the Philippines is *A. minimus flavirostris,* while *A. mangyanus, A. maculatus, A. litoralis,* and *A. balabacensis balabacensis* are other vectors.

Indonesia

Malaria was recognized as a serious public health problem in Indonesia a long time ago, with 30 million malaria cases and 120,000 deaths in 1955. However, the antimalarial campaign which started in 1956 as the national malaria control programme and changed in 1959 to the national malaria eradication programme has yielded satisfactory results, leading to a reduction of malaria cases to 117,000 in 1970. During 1960–1970, however, the programme was interrupted periodically due to political problems, shortage of funds, and unavailability of insecticides. This resulted in an increase in the number of malaria cases. According to WHO reports in the last few years, malaria cases are at about 150,000 a year. The overall slide positive rate (SPR) of malaria in Indonesia in 1979 was about 1.5%.

Falciparum malaria was found to be 36–46% of the total cases; many were resistant to chloroquine.

The principal mosquito vectors in Indonesia are *A. aconitus*, and *A. sundaicus*, while *A. subpictus, A. maculatus, A. nigerrimus, A. tesellatus* and *A. vagus* are other vectors in different islands in the country. Recently *A. aconitus* was found resistant to DDT in central Java. Other species are still susceptible to the insecticide.

Singapore

Singapore has been free from malaria since World War II. Malaria cases found in the hospitals in Singapore were imported from the offshore islands and neighbouring Malaysia, with 80–116 cases during 1978–1980. Surveillance of malaria cases in Singapore is strictly observed throughout the year.

Antimalaria Campaign

The main objectives of an antimalaria campaign should be to reduce mortality, morbidity, and prevalence of malaria in the country, so that it is no longer a problem. In the 1955–1970 era, we believed that malaria could be eliminated from the endemic areas through the national eradication programme with the aims of (1) eradicating malaria entirely from the country by DDT residual house spraying and the treatment of malaria cases detected actively and passively, and (2) preventing the reestablishment of endemicity of malaria in those eradicated areas. To fulfill these aims, two major activities were carried out, DDT house spraying and surveillance.

Spraying Operation

To kill adult *Anopheles* mosquito vectors of malaria in order to interrupt transmission, DDT residual house spraying is conducted in the endemic areas on one or two cycles annually. The one-cycle is applied in areas with a stable population and no apparent problems, just before the rainy season. The two-cycle spraying is used in communities with unstable population, at locations of high endemicity and problems in connection with the active transmission of malaria. The DDT spraying in the two-cycle spraying is made just 6 months after the first spraying or at the end of the rainy season. This operation is assessed from time to time by a malaria parasite survey on man and by entomological studies in the "index villages" to determine the prevalences of malaria and the vectorial capacity of the mosquitoes for longevity and density.

Surveillance Operation

Surveillance is carried out by case detection of malaria and treatment in the areas (1) where DDT spraying has been withdrawn due to success in terminating transmission, and (2) which are in the late attack phase with low endemicity.

Problems in Epidemiological Studies and Control of Malaria

The problems causing persistent transmission of malaria in the areas are as follows:

INTERNAL MIGRATION OF PEOPLE. The constant and frequent movement of large numbers of people from nonmalarious to malarious areas and vice versa is a regular feature of all Southeast Asian countries except Singapore, resulting in a persistent high endemicity of malaria in these areas. The population movement may be of a more or less permanent nature such as in land settlement areas, or more temporary ones including seasonal agricultural labour (e.g., sugar-cane plantations) or gem-mining. Such migration must be studied continuously with appropriate measures applied to minimize the problems in the malaria control programme.

MOVEMENT ACROSS NATIONAL BOUNDARIES. This practice is common among the people residing along the borders of the neighbouring countries in the environments of hilly, mountainous, and forested areas. Persistent malaria is reported among people in the communities in areas bordering Thailand with Malaysia, Burma, Laos, and Kampuchea. Moreover, in the last few years there has been an influx of many hundred thousands of Kampuchean refugees crossing the border to Thailand, including a considerable number of malaria cases, most of which are resistant to chloroquine and Fansidar. An antimalaria campaign has been carried out in the refugees camps and the surrounding Thai rural communities with successful results.

HABITS OF PEOPLE. In the Southeast Asian countries, rural people prefer to stay outdoors at their leisure time in early evenings because it is cooler and more pleasant. This is the place and the time that *Anopheles* vector mosquitoes of malaria are often active, and thus they may transmit malaria to man without being exposed to the DDT-sprayed walls of the houses.

FAILURE OF COMMUNITY PARTICIPATION. At present many householders are uncooperative in the DDT house spraying, especially after two or three cycles. They complain that DDT causes dirtiness, especially on the walls and under the roofs. DDT spraying is denied when there are sick people, especially children or women after childbirth, and religious ceremonies. In many cases only some parts of a house are allowed to be sprayed and not private rooms, sacred places, or rooms with many shopping materials. On many occasions, household owners are absent while working in the field; the houses cannot be sprayed. In Thailand complete house spraying was allowed in only 45% of the total households in 1980.

CONDITIONS OF TEMPORARY HUTS AND HOUSES. When the migrants come to settle in the village, they start to move into the deep forested lands and use them as newly opened for agricultural purposes. Small new farm huts and isolated dwellings are built in the remote areas previously uninhabited. Those are temporary in nature, roughly built without walls or rooms. Thus the house spraying yields negative results for such housing and the inhabitants are at great risk of contracting malaria.

DRUG-RESISTANT FALCIPARUM MALARIA. Today all countries in Southeast Asia have cases of chloroquine-resistant falciparum malaria, their prevalences and severity being variable and depending upon conditions in the respective countries. In countries such as Thailand and Kampuchea, malaria cases also resist Fansidar and a few cases have been reported to be resistant to quinine. These findings must be investigated and confirmed so that chemotherapy and chemoprophylaxis will be revised.

INCREASE IN VIVAX MALARIA. The relative prevalence of the species of malaria parasites, i.e., *P. falciparum, P. vivax,* and *P. malaria,* has shown wide variations from country to country; within each country; and year by year due to control and eradication programmes. Except for Indonesia, *P. falciparum* is the dominant species in the region, its prevalence varying from 53% in Vietnam to 77% in Burma. In the last few years vivax malaria has been increasing in prevalence, resulting in chronic infections.

CHEMOTHERAPY. Chemoprophylaxis is essential for the armed forces personnel and migrant workers entering newly opened agricultural lands which are endemic areas of malaria. However, it is difficult to get those people to take antimalarial drugs at the right time and in correct dosages unless administered in person by the antimalaria staff.

INSECTICIDE-RESISTANCE OF VECTOR MOSQUITOES. It is necessary to test the sensitivity of vector mosquitoes to DDT, since in some countries a few *Anopheles* develop resistance to DDT insecticide.

CHANGE OF HABIT OF VECTOR MOSQUITOES. The principle of DDT residual house spraying is to kill *Anopheles* adult mosquitoes by knowing their habits of biting and resting indoors. If the mosquitoes change their habits to be exophilic and exophagic, however, the house spraying with DDT would not affect them. We always have to study the biting habits of local vectors of malaria in the area in order to evaluate the results of spraying. In Thailand *A. minimus* and *A. balabacensis* in some areas have changed their habits to bite and rest outdoors more than indoors.

FINANCE. A complete understanding of the financial implications and commitments of the programme and full backing and support from the governments, the politicians, and people are necessary before embarking on a prolonged country-wide eradication programme.

Since there were many problems leading to unsatisfactory results in 1975–1976, WHO and the Southeast Asian developing countries agreed that the antimalaria campaign should be changed from malaria eradication to malaria control in the endemic areas.

Amoebiasis

Amoebiasis has a worldwide distribution and is most prevalent in the tropics, especially in communities with poor sanitation and low standards of personal hygiene, urban slums, and also in isolated units lacking medical care such as mental institutions or orphanages. The disease exists in an endemic form in the tropics. Many outbreaks are due to a water supply heavily contaminated with *Entamoeba histolytica* infected faeces with seepage between water and sewer pipes.

The prevalence of intestinal amoebiasis in the Southeast Asian countries have been found in general to be moderate, e.g., 6–13% in Indonesia, 2–8% in the Philippines, 2–6% in Thailand, and 4–8% in Malaysia. The incidence is higher in the communities where poor people live in crowded, unhygienic, and unsanitary areas such as slums.

In Thailand, the overall incidence of intestinal amoebiasis is rather low. In many provincial areas the infection rates of *E. histolytica* ranged from 0.5% to 3.5% in 1970. The incidence in patients seen at the outpatient department of a larger medical school hospital in Bangkok was about 10.7% in 1979.

Treatment

There are three new effective drugs in the treatment of intestinal amoebiasis in our region, metronidazole (Flagyl), tinidazole (Fasigyn), and ornidazole (Tiberal). The first drug is given at a dosage of 750 mg (3 tablets) 3 times a day for 5–10 days; the second drug, at a single dose of 2.0

gm (4 tablets) for 3 days; and the third at the dosage of 500 mg (1 tablet) 3 times a day for 5–10 days.

Amoebic Hepatitis and Amoebic Liver Abscess

Amoebic liver abscesses are found to be prevalent in many provinces of Central Thailand, where intestinal amoebic infection rates are about 5–8%. There were 64 cases admitted to Siriraj Medical School Hospital in Bangkok in 1976.

Giardia lamblia, an intestinal flagellate protozoa, may be found in the duodenum, upper jejunum, and occasionally gall bladder of man. Particularly in preschool and school children, the parasite invades the mucosa of the intestine resulting in clinical manifestations. In heavy infections, diarrhea is the main symptom. The stools are light-colored, containing excessive amounts of fat with special odor. The fecal examinations usually show the cystic stage of the parasite. Trophozoites are rarely present in stool, except in cases with severe diarrhea. The overall prevalence among children in rural communities in Thailand ranges from 10% to 20%.

Atabrine hydrochloride, tinidazole, and ornidazole are effective drugs in the treatment of giardiasis.

Soil-Transmitted Helminthiasis

The important soil-transmitted helminths causing diseases in the countries in Southeast Asia are *Necator americanus, Ancylostoma duodenale,* for hookworm; *Ascaris, Trichuris,* and *Strongyloides.* Hookworm infection is widespread among the rural population, since about 80% of the people in these countries earn their living by agriculture, which includes rice and various agricultural crops. This facilitates the transmission of infection. The estimated number of hookworm cases in Southeast Asia is not less than 80 million. Ascariasis is higher among children with poor hygiene and sanitation; the prevalence is about equal to that of hookworm infection. Trichuriasis is prevalent in the warm and moist climate with abundant rainfall such as in Peninsula Malaysia and the island areas of Indonesia. *Strongyloides* has the lowest prevalence of the soil-transmitted helminths.

Hookworm Disease

Ninety percent of the hookworm in Southeast Asia is caused by *Necator americanus,* with the remainder by *Ancylostoma duodenale.*

Recently *A. ceylonicum* has been reported to be the third species of human hookworm infection in Thailand. The clinical features of hookworm disease commence when a large number of hookworms are localized in the small intestine, resulting in blood and protein loss from the

host. The common symptoms are fatigue and weakness, dyspnoea, palpi-
tation, pallor, and epigastric fullness. In heavy infection of long duration,
the patient becomes markedly anaemic with mental and physical retardation. Extensive oedema and cardiac enlargement with malnutrition are the
late results, and death may occur if treatment is not administered. The
prevalence of this infection in the Southeast Asian countries range from
10% to 75%.

Socioeconomic and environmental factors are responsible for the prevalence of hookworm infection. The socioeconomic factors are (1) low
educational level resulting in poor personal hygiene and sanitation, (2)
unavailability of hygienic latrines in individual houses, (3) defaecating on
the ground in the bush near the houses of the villagers, (4) habit of walk-
ing barefoot by the rural people, thus easily acquiring the hookworm
infection, and (5) certain occupational habits of the people in relation to
frequent contact with the soil, e.g., rubber tapping early in the morning in
rubber plantations and tin-mining in south Thailand and Malaysia.

The environmental factors include (1) poor environmental sanitation
due to inadequate water supply and improper disposal of human excreta,
(2) moist climate and soil, favorable for the growth of the eggs into infective larvae, and (3) type of soil which is shaded, loose, moist, loamy, and
sandy, thus favorable for the growth of the parasite.

Ascariasis

This infection with *Ascaris lumbricoides* is more frequent in children than
in adults, since it is associated with poor personal hygiene and sanitation.
A single worm of ascariasis may cause symptoms if it undergoes "ectopic
migration" resulting in jaundice due to obstruction of the bile duct,
appendicitis or perforation of the bowel. The overall presence of infection
in each country in Southeast Asia ranges from 10% to 80%.

The mode of infection is by the ingestion of young infective eggs in the
soil which reach the mouth by hands or dirty foods. Infected children who
fail to use sanitary toilet facilities contaminate the playground with their
excreta and thus seed the soil with eggs of the parasite. This provides a
source of new infection for other children and of reinfection for
themselves.

Trichuriasis

While a small number of *Trichuris trichuria* produces no symptoms, a
large worm burden in children and adults may be associated with intermittent diarrhoea for a long duration, bloody and mucus stools, abdominal
pain, anaemia, and weight loss. Occasionally these children have prolapse
of rectum with visible adult worms on the mucosa of the bowel. The prevalence of infection in many islands of Indonesia, Philippines, and Malaysia
ranges from 25% to 90%.

Strongyloidiasis

Strongyloides stercoralis may occasionally cause gastrointestinal disturbances including abdominal pain, nausea, vomiting, diarrhoea, anorexia,
and generalized urticaria. The frequency of strongyloides in Southeast
Asia is reported as 0.1–8.0%.

Control Measures for
Soil-Transmitted Helminthiasis

The standard measures consist of (1) proper disposal of human faeces and
prevention of soil pollution by hygienic latrines, (2) mass treatment of the
population to reduce the reservoirs of infection, and (3) health education
to the community for better personal hygiene and environmental
sanitation.

These measures, however, are at times impractical for such reasons as
resistance to changing habits, economic status of the villagers, cultural status, and failure to obey. Therefore, it is necessary to study the ecology of
the indivdual diseases and the socioeconomic status of the community
before prevention and control measures can be applied.

Three major problems relate to the effectiveness of control measures.
First, hygienic latrines may be impractical for construction in isolated rural
areas due to lack of money and of water supply. Furthermore, villagers may
continue to defecate in the bush, while reserving their new latrines for
guests. Second, mass treatment of the population—pyrantel pamoate and
mebendazole for hookworm and ascariasis; mebendazole for trichuriasis;
strongyloides with thiabendazole and pyrvinium pamoate—may be impossible. Third, health education leads to the successful prophylaxis of these
infections. Epidemiology and the necessity for the prevention of soil pollution are key factors.

Capillariasis

Capillariasis, a new chronic disease in Southeast Asia, is caused by an
intestinal nematode *Capillaria philippinensis.* Manifestations include
diarrhoea, abdominal pain, cachexia, muscular wasting, edema, electrolyte
depletion, and malnutrition. Progressive loss of weight and marked protein and electrolyte depletion may lead to death within 3–4 months. This
disease is endemic in some provinces in the northern part of the Philippines and in many in northeast and central Thailand. A total of more than
2,000 cases were reported in these two countries.

The causative agent is a tiny and slender nematode, about 2–4 mm in
size, inhabiting the small intestine, especially the jejunum. Eggs detected
in the faeces confirm the diagnosis. The disease occurs in the communities
where people customarily eat raw freshwater fish. The natural life cycle of
this nematode is between freshwater fish and fish-eating birds; when man
eats the fish, he becomes infected.

The drug of choice is mebendazole, 400 mg daily for 15–20 days. The gastrointestinal symptoms are improved within 3–4 days and the cure rate is 90–100%.

Filariasis

Filariasis in Southeast Asia is caused by *Brugia malayi*, so called malayan filariasis, and by *Wuchereria bancrofti*. The parasites are nematodes localized in the lymphatic system of man. The signs and symptoms include lymphangitis, lymphadenitis, orchitis, epididymitis, hypertrophy and hyperplasia of lympatics, progressive obstruction of lymphatic channels by fibrous tissue, and elephantiasis of the affected organs. Elephantiasis of the legs and arms is the main characteristic of malayan filariasis, while involvement of the genital organs is more commonly seen in bancroftian filariasis. The parent worms produce microfilariae which enter the circulating blood and appear in the peripheral blood at intervals usually during the night. These microfilariae do not contribute to the pathological changes or cause the symptoms as mentioned above. They are observed as means of diagnosis of the infection. Many mosquitoes are filariasis vectors including *Culex, Anopheles, Mansonia,* and *Aedes,* according to the local environment.

Filariasis is found with varying endemicity in Burma, Thailand, Malaysia, the Philippines, and Indonesia.

MALAYAN FILARIASIS. Malayan filariasis has an endemic character of small localized foci at village level distributed here and there in rural areas. It can be found in co-endemicity with bancroftian filariasis. Also a mixed infection of malayan and bancroftian filariasis in the same individual is not uncommon in some particular areas.

There are two types of this infection: nocturnally periodic is found in five countries while nocturnally subperiodic is limited to forested areas in Malaysia and the Philippines. The subperiodic form has many animal reservoirs including cats, monkeys, and some wild animals. The mosquito vectors for nocturnally periodic filariasis include *Mansonia uniformia/ indiana* and some of the *Anopheles* species. For the subperiodic, transmission is by swamp forest mosquitoes, *Mansonia dives/bonneae.* Many aquatic plants and long-leaf grasses growing in swampy waters serve as the breeding sites for *Mansonia.*

The infection rate ranges form 10.5% to more than 25% according to the endemicity of the area. Elephantiasis is found in more females than males; it usually begins at 20–40 years of age. Malayan filariasis is not a cause of mortality.

BANCROFT FILARIASIS. This disease is widely distributed in nearly all countries of Southeast Asia, with clinical manifestations resembling those of the Malayan form. The urban type is carried by *Culex fatigans* in Burma and Indonesia, the rural variety by *Anopheles* and *Aedes* in the Philippines and Thailand. The *Anopheles* are *maculatus* and *whartoni* in Malaysia, with *minimus flavirostris* in the Philippines. *Aedes poecilus* is another Philippine vector, and *Aedes niveus* appears in Thailand.

Recent reports describe a new form of filariasis, caused by *Brugia timori* on the island of Timor in eastern Indonesia. It is carried by *Anopheles barbirostris*.

CONTROL MEASURES. There are two procedures in the control of filariasis in Southeast Asia: (1) to kill microfilariae of *B. malayi* and *W. bancrofti* by chemotherapy, and (2) to reduce vector mosquitoes by insecticides, physical means, or other relevant methods.

To Kill Microfilariae by Chemotherapy. This is the standard method in any country. The aim is to reduce the number of microfilariae in the peripheral blood so that transmission would be arrested. Diethylcarbamazine is the only drug used for this purpose, either by mass therapy in the high and moderate endemic areas or by case detection and individual treatment in the areas of low endemicity. Diethylcarbamazine is also administered in cooking salt, food, or drink. Many reports showed successful results after using this approach to administration.

Measures Against Mosquito Vectors. Those methods include (1) insecticides to kill adult mosquitoes (DDT house spraying) and larvae (application of insecticides into water in the breeding places of mosquitoes), (2) physical measures to clear mosquito breeding places, and (3) biological approaches to vector control (studies in laboratories and then application of the methods in the field). Satisfactory results depend upon local conditions, environment, and the level of community participation in endemic areas.

Angiostrongyliasis

Angiostrongylus cantonensis is an animal nematode parasite invading the pulmonary blood vessels of rats and other rodents and causing eosinophilic meningitis or meningo-encephalitis in man. The disease is regarded as an emerging zoonosis causing public health problems in some rural communities.* The *Angiostrongylus* adult worms in the lung of rats produce eggs and larvae which are evacuated in their faeces through pulmonary alveoli, bronchi, trachea, and oesophagus to the alimentary tract. Susceptible snails and slugs including *Pila* and *Achatina* ingest the rats' faeces and the third-stage infective larvae develop. The complete cycle from mollusk to rodents is by three routes: (1) rodents ingest infected mollusk tissue, (2) infective larvae from mollusk-contaminated drinking water are ingested by the rodents, or (3) mollusk tissue is ingested by freshwater prawns, land crabs, or planarians which act as carrier hosts, after

---

*Zoonosis—an infection or infestation shared in nature by man and lower vertebrate animals.

which rodents ingest these carriers. In the rodents, the infective larvae develop in the brain and then migrate to the pulmonary arteries.

Man contracts this infection by eating raw snails containing larvae of *A. cantonensis*, drinking contaminated water with infective larvae, or ingesting carrier hosts. In man the immature worms migrate from the alimentary tract to the brain, where they die. When the worms migrate to the central nervous system, however, they cause meningitis or meningo-encephalitis with high eosinophilia in the blood. Clinically, patients suffer severe headache at times with confusion, disorientation, and incoherence. Other symptoms include vomiting, cervical rigidity, impairment of vision sometimes leading to blindness, and drowsiness leading to coma. After 1–3 weeks of these signs and symptoms, the patient recovers as he develops resistance to the parasite.

Ocular angiostrongyliasis may occur due to migration of the immature worms; this may result in visual impairment.

Laboratory findings include eosinophilia in the blood and cerebrospinal fluid; the larvae may be isolated from the spinal fluid. Haemagglutination and other serologic tests are useful in diagnosis of the disease.

EOSINOPHILIC MENINGITIS AND MENINGO-ENCEPHALITIS. *Angiostrongylus cantonensis* is considered to be the most important cause of eosinophilic meningitis and meningo-encephalitis in man in Southeast Asia. Hundreds of cases have been recognized each year. Cases of eosinophilic meningitis and meningo-encephalitis have been reported in many provinces of northeast Thailand, where the people eat raw snails (and raw fish). In 1965–1966 more than 200 cases were recorded in three provincial hospitals in northeast Thailand. Nearly all had a history of eating raw snails prior to the occurrence of the symptoms, and severe headache was the main complaint. The disease had a seasonal pattern, peaking in July through September, which is the rainy season in Thailand producing numerous edible snails.

Six species of rodents in Thailand were found to harbour *A. cantonensis* in most of Thailand. Thirteen species of edible snails were reported to be infected with *A. cantonensis* larvae in many areas of the country, and *Pila* snails seemed to be the most active in transmission of the disease.

In Indonesia, many cases of eosinophilic meningitis due to angiostrongyliasis have been reported in North Sumatra and one definitely confirmed case in Central Java, while *Angiostrongylus cantonensis* has been found prevalent in nine species of rodents and three species of mollusks in four major islands of Indonesia: Sumatra, Java, Sulawesi, and Flores.

In the Philippines, *Angiostrongylus cantonensis* worms were found in six species of rats, while infective larvae were recovered in seven species of local mollusks. Although many human cases of eosinophilic meningitis and meningo-encephalitis have been reported in some areas of the country, none of the cases has been verified.

In Vietnam and Laos there are many reports of meningo-encephalitis and ocular angiostrongyliasis caused by *A. cantonensis*, while one species of rats and one species of mollusk were found to harbour *A. cantonensis* worms.

There is no specific chemotherapy for this disease.

Trichinosis

Trichinosis is a food-transmitted parasitic zoonotic infection caused by *Trichinella spiralis*. The disease manifests itself with occasional outbreaks in communities in which raw pork or improperly cooked pork food is usually eaten. In Southeast Asia cases of trichinosis are reported from Thailand, and less in Burma, Indonesia, and Indochina countries. In Thailand in the last 20 years, 1962–1981, there were 47 outbreaks of human trichinosis involving more than 2,316 patients with 1.6–6.4% death rate. The mortality rates were reduced a great deal during the last 10 years due to better clinical management and very effective chemotherapy.

The animal reservoirs of trichinosis in Thailand are usually hill-tribe pigs, wild boars, and wild rats, and the infection is presumably enzootic. The worms are located in muscular tissues as encysted larvae. After man ingests raw infected muscles of hogs, the worms are freed from their cysts, enter the small intestine of man, developing to mature worms and producing more larvae, invade the intestinal mucosa, and enter the lymphatics and blood stream. They then encyst in the skeletal muscles of man.

The manifestations include nausea, vomiting, fever, abdominal pain, diarrhea, eosinophilia (usually 20–40%), and leucocytosis, lasting for 5–7 days. Myalgia and myositis with high fever may occur many days later with periorbital edema. Death in severe cases is by pneumonia, secondary infection, or cardiac failure. In most cases the symptoms improve slowly with some muscular pain persisting for a few months. Encysted larvae of *Trichinella* may be recovered by biopsy of the muscles, especially in the calf of the leg.

In North Thailand since 1962 outbreaks of trichinosis have been reported up to four times annually. Diagnosis is confirmed by a skin test using commercial antigen and biopsy to recover *Trichinella* encysted larvae in the late stage of the disease.

Treatment includes massive doses of corticosteroids. Thiabendazole in a dosage of 50 mg per kilogram body weight per day for 5–7 days is effective against the migrating and encysted larvae. Mebendazole also gives promising results.

Gnathostomiasis

Gnathostomiasis or creeping swelling in man is a zoonotic disease caused by immature *Gnathostoma spinigerum*, a nematode worm normally found in domestic cats, dogs, and other carnivorous animals in raw fish-eating areas.

The adult worms of *G. spinigerum* coil in a tumor-like mass in the wall of the stomach of the animal reservoir hosts. Eggs pass in the faeces to become larvae in the water where they are ingested by cyclops. The infected cyclops are then eaten by fish, frogs, snakes, fowl, etc., which produce the third-stage infective larvae of the worms encapsulated in the flesh of the animal hosts. The life cycle is complete when a cat ingests infected raw flesh; the worms become localized in the stomach wall and mature after about 6–7 months. Man acquires infection with *G. spinigerum* by ingestion of the hosts containing infective third-stage encapsulated larvae. The immature worms then migrate in subcutaneous tissue and other tissues and organs resulting in migratory swelling. The worm may invade the central nervous system, producing a severe headache, paresis, paralysis, coma, and at times, death. *Gnathostoma*, as well as *Angiostrongylus*, produces many cases of meningo-encephalitis with eosinophilia. Also a number of cases of ocular damage have been reported. Diagnosis of the disease is by clinical features and high eosinophilia (10–90%). It is confirmed by skin test using crude antigen.

There is potential infection in man in Indonesia, Malaysia, the Philippines, and Vietnam since there have been reports of adult worms in animals in those countries.

In Thailand gnathostomiasis has been found in a few provinces in the central part of the country, where the local people have habits of eating special kinds of raw fermented fish food such as "Som-Fak" and "Pla-Som." Cases with the symptoms of the central nervous system involvement have also been reported from northeast Thailand. During 1961–1963 about 900 cases of suspected gnathostomiasis were admitted into 92 provincial and Bangkok hospitals.

Since cats and dogs in Thailand are considered to be important animal reservoirs for gnathostomiasis, surveys on the infection in these animals were made during 1965–1970. Four percent of domestic cats in Bangkok were found to harbor *Gnathostoma* and 1.1% of dogs from the Rabies Prevention Centre in Bangkok.

Four species of cyclops are the first intermediate hosts while 30 species of fishes and vertebrates are the second intermediate hosts, i.e., 2 species of freshwater fish, 3 species of amphibians, 5 species of reptiles, 3 species of fowl, 2 species of crabs, and 13 species of rodents and monkeys.

There is no specific drug for this parasitic infection. The surgical removal of the worm may be tried when it is located superficially. Thiabendazole has been tried at a dosage of 2 g daily for 20 days with some effective results.

Schistosomiasis

Schistosomiasis in Southeast Asia may be divided into three etiological groups: the first, *Schistosoma japonicum* with *Oncomelania* as snail vector on the island countries including the Philippines and Indonesia; the second group, *S. mekongi* with *Tricula aperta* as vector snail found in Laos and Kampuchea; and the third group, caused by *S. japonicum*-like worms with unknown snail vector, reported in Thailand and Malaysia.

SCHISTOSOMA JAPONICUM INFECTION.

Philippines. The infection is caused by *S. japonicum* with *Oncomelania quadrasi* as the snail vector. The endemic areas cover 22 provinces in the southeastern part of the country (6–14° North latitude) with more than 700,000 known cases in an exposed population of 4 million. Most cases are found in the Islands of Leyte, Samar, Davao, and Mindanao. The disease in the rural areas occurs in farmers and their families, especially children and young adults aged 20–24 years. The infection is also found in freshwater fishermen. There are many species of animals serving as reservoir hosts of schistosomiasis in the Philippines.

The National Schistosomiasis Control Commission has been established in the Philippines to cope with the problem. The control programme includes clinical studies and chemotherapy with Praziquantel (heterocyclic pyrazinoisoquinoline). Vector snail control, improvement of environmental sanitation, and health education are also pursued.

Indonesia. S. japonicum in Indonesia is localized in small areas around Lake Lindu, and in Napu Valley (50 km southeast of Lake Lindu) on Celebes Island. The snail vector, *O. hupensis lindoensis*, is widely distributed in the endemic areas. The reservoir animals include wild deer, dogs, wild pigs, cattle and many species of rats.

SCHISTOSOMA MEKONGI. Schistosomiasis caused by *S. mekongi* is endemically on Khong Island in the Mekong River in Southern Laos and in many areas in northern and central Kampuchea (Cambodia). The snail vector has recently been found to be *T. aperta*, an aquatic snail in the Mekong River and its tributaries. Dogs are the sole reservoir host.

Laos. The first case of schistosomiasis in Laos appeared in an 18-year-old Lao-Eurasian student in Paris who had spent his early life on Khong Island in the Mekong River in Southern Laos. In this case numerous *S. japonicum*-like eggs were found on liver biopsy. Again during 1966 Barbier reported four cases of schistosomiasis among Laotians admitted into a hospital in Paris. These victims had also resided on Khong Island. A WHO survey team in 1967 reported that Khong Island had been verified as an endemic area of a *S. japonicum*-like infection with a rate of 11.9% as proven by skin tests and stool examinations.

Another study on Khong Island was carried out by the Faculty of Tropical Medicine, Mahidol University in 1969. The procedures included skin tests, stool examinations, and miracidial hatching; they found an incidence of 14.4%. Most cases were youth ranging in age from 4 to 27 years, with a ratio of males to females of 2:1. The clinical signs in most cases were intermittent diarrhea, with a full-blown dysentery in a few. Up to 56% showed hepatomegaly and splenomegaly with 4% developing ascites. Among the local canines, 11% harbored these schistosomes; the parasite could not be found in any other domestic or wild animals. The snail vector, *T. aperta*, also known as Lithoglyphosis aperta, showed an infection rate of 0.3%.

Studies during 1970–1973 proved that the vector snail had three forms, termed alpha, beta, and gamma, all of which are aquatic in nature with an adult size of 3 × 2 mm. These snails usually attached themselves to sticks or stones in slow-running and shallow water.

We are continuing surveillance in the bordering Thailand areas, to prevent spread of this infection into our country.

Beginning in 1968, schistosomiasis in Kampuchea was first identified at Kratie town on the Mekong, in the families of fishermen living in floating rafts on the riverfront. It is reported to be identical with the Khong Island type with an incidence of 7–10%, predominantly in children aged 1–14 years. The disease has also been identified among refugees from Kampuchea fleeing into Thailand; this is also probably *S. mekongi*.

INFECTION WITH A *S. JAPONICUM*-LIKE PARASITE.

Thailand. The first cases were reported from two districts of Nakhon-Srithammaraj Province, southern Thailand with a total of 50 diagnosed by rectal biopsies. A majority were 21–40 years of age with only mild gastrointestinal disturbances. Despite several years of investigations, neither a vector snail nor animal host could be identified. The infection in this area disappeared. During 1964–1973, cases of *S. japonicum*-like infections were reported from north Thailand.

Malaysia. In 1975, nine cases of *S. japonicum* were reported among 231 autopsied aborigines who had died from a variety of diseases over a period of eight years in the states of Pahang and Perak. In a 1978 study, nine schistosome infected persons were identified, but it was not possible to find endemic foci.

Paragonomiasis

Paragonomiasis, lung fluke infection, is prevalent in Thailand, the Philippines, Laos, Vietnam, and occasionally in Malaysia and Indonesia. Two species are known to be infective to man: *Paragonimus westermani* causes chronic coughing, hemoptysis, and cerebral manifestations. *P. heterotrema* produces only migratory discomfort. Four other species of *Paragonimus* are found in cats, dogs, rats and other rodents, mongoose and tigers.

In human cases, the parasite is encapsulated in lung tissue, producing eggs which pass upwards in the respiratory tract to the mouth or down the intestinal tract with the faeces.

The snail vectors include *Brotia asperata* in the Philippines; *B. costula* in Malaysia; and probably *Melanoides tuberculata* in Thailand. When the cercariae leave these snail vectors, they enter many types of mountainous crabs as teh second intermediate host. The vector crabs in Thailand are *Potamon smithianus*, *Parathelphusa degasti*, and *Parathelphusa sp.* The infection rates in crabs in the endemic areas range as high as 50%.

Man contracts this infection by eating raw crabs; this is a common practice among rural people. The cerebral signs and symptoms may become quite severe, with convulsions and hemiplegia followed by coma and demise.

The *Paragonimus* cysts are easily identified on chest X-rays; skin tests are useful for screening. Bithionol is the specific drug for curative therapy.

Opisthorchiasis

This liver fluke infection in Southeast Asia is caused by *Opisthorchis viverrini*; it is estimated that 5–6 million people are suffering from it in Thailand with 1 million in Laos. Scattered cases are reported from west Malaysia. The infection rate varies from 10 to 90%. Opisthorchiasis is not only a medical and public health problem, but also a serious economic impediment.

The adult worms vary from 5.5 to 9.5 mm in length and 0.8 to 1.7 mm in width, with the egg about 16–29 microns in size. An adult worm produces as many as 1,100–2,400 eggs per day, which are discharged into the bile ducts and passed in the faeces. The life cycle follows, man–freshwater snail–freshwater snail–freshwater fish–man.

While man is the definitive host, dogs and cats are reservoirs. *Bithynia goniomphalus*, a freshwater snail, is the host in northeast Thailand; *B. funiculatus* in the north; and *B. laevis* in central Thailand.

After *Opisthorchis* cercariae leave the vector snails, they enter the second intermediate host, cyprinoid fish, *Cyclocheilicthus siaja*, *Hampala dispar*, and *Puntius orphoides*. In high endemic areas, the infection rates in these fish are 50–90% with each fish harboring 20–50 metacercariae. Humans contract the infection by eating these fish.

The high incidence of this disease is attributed to comsumption of raw fish, lack of latrines so that faeces with the parasite are in the bushes, and the prevalence of vector snails and cyprinoid fish.

In the human disease, adult worms inhabit the biliary tracts to cause adenomatous hyperplasia of the biliary epithelium.

Diarrhea, hepatomegaly, and jaundice are clinical manifestations with portal cirrhosis and cholangiocarcinoma as late developments. The diagnosis of this liver fluke infection is usually made by demonstration of the eggs in human faeces.

Praziquantel, the drug of choice in schistosomiasis, is 100% effective in the treatment of opisthorchiasis.

Fasciolopsiasis

A giant intestinal fluke, *Fasciolopsis buski*, the etiological agent in this disease, establishes itself in the duodenum and jejunum of man and pigs to produce abdominal pain, diarrhea, and malnutrition. This infection is active in specific areas of Thailand, Vietnam, Cambodia, Laos, and Burma. In one district of Central Thailand, the incidence ranges from 3.3 to 38.5%. Death may occur from cachexia or bacterial infection, especially in children.

Two species of freshwater snails, *Polypylis hemisphaerula* and *Trochor-bis trochoides*, are the first intermediate host in which cercariae develop. When they become mature, the cercariae leave the snails for encystment in a variety of species of water plants. Man and pigs acquire this infection by eating these plants.

Hexylresorcinol and tetrachloroethylene are the effective drugs in expelling the worms.

Echinostomiasis

With *Echinostoma malayanum* as the causative agent in Thailand, Laos, Malaysia, and Indonesia, the overall incidence ranges from 0.5 to 10%, but with a higher prevalence in residents of rural areas who eat raw snails and tadpoles, which harbor the metacercariae.

The clinical manifestations include abdominal distress and diarrhea. Tetrachloroethylene and hexylresorcinol are highly effective therapeutic agents.

Taeniasis

Taenia saginata infection is prevalent in all countries of Southeast Asia, primarily among communities whose residents eat raw or inadequately cooked beef. The adult cestode worm, about 12 mm in width and 5–10 m in length, attaches its scolex, or head, to the mucosa of the ileum with its immature proglottid along the gut wall. The eggs from the embryonal parasites are passed through the anus and devoured by cattle and buffaloes with penetration of the gut to the peripheral tissues and musculature.

Niclosamide (Yomesan), mebendazole (Fugacar), and quinacrine (Atabrine) are the drugs of choice, and completely effective.

The cysticerosis stage of *Taenia solium*, usually found in pigs, also occur in man. Small cysts appear in muscular and subcutaneous tissue throughout the body, but with only limited signs and symptoms. If these cysts appeared in the brain, however, nervous manifestations appear, including headache, convulsions, and paralysis. The effects of drug therapy are limited.

Summary

The rural communities of nine countries in Southeast Asia including Burma, Thailand, Vietnam, Laos, Kampuchea, Malaysia, Singapore, the Philippines, and Indonesia have been reported to be suffering from 17 local endemic parasitic diseases of public health importance. These are malaria, intestinal amoebiasis and amoebic liver abscess, giardiasis, hookworm disease, ascariasis, trichuriasis, strongyloidiasis, capillariasis, malayan and bancroftian filariasis, angiostrongyliasis, trichinosis, gnathostomiasis, schistosomiasis, paragonimiasis, opisthorchiasis, intestinal fluke infections, and taeniasis. The epidemiology of these 17 parasitic diseases together with short accounts for their treatment, prevention, and control has been described. These diseases are the basis of economic distress to the affected communities, due to morbidity and mortality, and the loss of workmen resulting in a diminishing national income through the decline in agricultural productivity. Harinasuta declared that the individual countries of Southeast Asia must assume responsibility to eradicate or control these diseases for overall welfare and progress.

Discussion

WILLIAMS: I have no doubt that parasites are very important in the developed areas, but I would like to ask Warren whether he thinks he has made the case that parasites are more important than the *E. coli*, which lives in the intestine, just as do many parasites. Sometimes the *E. coli* causes other disorders, however. Do you think that more than part of the flora of man sometimes becomes pathological? Can you make the case more strongly than that?

WARREN: You are talking about problems of the North?

WILLIAMS: Yes, I am not questioning the problems in the South.

WARREN: It is not a major problem, but it would be significant even if parasitic diseases didn't exist in the South.

Given the prevalence of many of these diseases, one big problem is an infection such as trichomonas, which is annoying but not life-threatening. An enormous number of women who have trichomonas are being treated with metronidazole. There is some question about whether it is mutagenic, and possibly carcinogenic in animals. Given the possibility that metronidazole may be a significantly carcinogenic drug, we don't like to use it even for treating patients with amebiasis, which is a life-threatening illness.

The toxoplasmosis problem is significant with respect to the danger for congenital disease and ocular disease, and so on.

Infections with pneumocystis, strongyloidiasis, and toxoplasmosis are appearing in people who are being treated with anticancer drugs and immunosuppressant drugs.

WILLIAMS: They aren't of relative importance. Do the people from Europe feel the situation is similar to that in the United States, or of lesser or greater importance?

CAPRON: The picture is much the same, except you did not mention the problem of ascariasis.

WARREN: That is rare.

CAPRON: In humans it is a rather important problem, with new cases being observed each year.

I would like to put more emphasis on the diseases of European countries, where in the South, again due to the foci of transmission, we have a very serious problem with older migrant workers from Turkey and Portugal, for example.

I would like to endorse what Warren said about toxoplasmosis in Europe as a public health problem.

SELL: I would like to make two points. First, it may be a tactical error to try to justify research in parasitology based on diseases in the North. Emphasizing the importance of these diseases invites a comparison with other diseases in the North.

My second point is that one often finds that a few of these diseases are significant in immunosuppressed patients. The pneumocystis, for example, is a common complication in such patients, where toxoplasmosis, which infects at least 40% of the population in the eastern United States, is an insignificant disease in immunosuppressed patients. That is an interesting scientific fact, and one that needs to be pursued rigorously. But it does tend to blunt the argument about the significance of these diseases in immunosuppressed patients.

WARREN: As a member of the Infectious Disease Society of America I am disgusted with the attitude of infectious disease researchers in the United States. What is the major concern of the American pharmacological industry? A concern about patients in tertiary care facilities who are immunosuppressed, who are developing opportunistic infections.

The greatest single concern at Infectious Disease Society meetings is these odd bacterial infections that occur in patients with terminal disease, or fungal infections.

BOOTH: You don't know how lucky you are in the United States to have infectious disease physicians. Most other countries don't have them because in many parts of the developed world people say there is no need for infectious disease physicians, the generalists can take care of any problems.

If I may speak for Britain, the present state of research on infectious diseases is woefully inadequate in terms of deep specialty interest and laboratory back-up. The successful laboratory integration in departments such as that at Tufts, where there is a top-grade laboratory organization and clinical structure, is a really good setup for looking at both parasites and bacteria.

That development has not happened in Britain and certain other Western European countries, where microbiology tends to be found in separate departments of microbiology. To some extent one of the defects of separate departments of parasitology is that they do not cover the clinical aspects.

So while you may find the situation unsatisfactory in the United States, it is far better than in other countries.

DAVID: When you said malaria was important, over a 6-week period in Boston I was surprised to find 14 cases of malaria, two of them premature American infants who had to be given blood transfusions; the others came from Asia.

The problem there is education. Physicians are taking a long time to make a diagnosis, and thus there is a delay in implementing treatment.

NELSON: I would like to comment on the benefits that can be given to people in the developing world, which has the largest number of parasitic diseases, by researchers in the developed countries who are working on the same parasites.

If one looks at fundamental research in relation to the biology of parasites, the transmission of parasites, and the chemotherapy, a lot of it is being done by people who may think it a trivial public health problem, whereas it is a very serious infectious disease problem in developed countries. Major advances have been made because these parasites exist in the North.

3

Human Helminthic and Protozoan Infections in the North

Julia Walsh

Helminthic and protozoan diseases stand out as uniquely characteristic of the tropics of the South. While these parasites may be more common in tropical climates and the intensity of infection may be greater, a surprisingly large number of individuals are infected in the temperate North.[1] Furthermore, it should be noted that most of the research on parasites, their life cycles, and the means of treating and controlling them has been performed either in Europe or America or by investigators from these areas. Although much of this research was done for humanitarian or scientific reasons, the justification for teaching and investigation has usually been on a more practical level: to safeguard the health of travelers abroad. Even more important has been the impetus provided in the last several decades by the military incursions into the tropics.

But with respect to medical parasitology, major justifications for a reasonable amount of teaching time and for further research exist directly within the North. Dr. Eli Chernin has observed, "The fact is that some protozoan and helminth infections are being exchanged at every level of our affluent society almost with the regularity of stock market transactions."[2] Dr. Franklin Neva has stated, "The North American economy does maintain a modest supply of parasites pathogenic for man. The United States citizen can acquire amebiasis, giardiasis, pinworms, and strongyloidiasis, for example, without so much as a passport application."[3] It appears that there is far more than a modest supply of worms and protozoa available on the U.S. market. A reasonably conservative estimate, based on surveys in the last 10 years, is that there are tens of millions of helminthic infections in the continental United States and many more protozoan infections. Although most of these are benign, they do cause some physical and much psychological misery, as well as a substantial cost to the health care system. Furthermore, they may lead to disabling and even fatal illness, for example *Ascaris* obstruction of the bile duct, severe trichinosis, disseminated strongyloidiasis in immunosuppressed patients, mental retardation from congenital toxoplasmosis, and liver abscess in amebiasis. The National Reference Centre for Parasitology in Canada estimated the cost for hospitalization for parasitic diseases in 1973 to be $4 million (20,000 days of hospitalization at $200 per day).[4] The United States has 10 times the population but a similar prevalence of parasitic infestations, and the daily bed cost for hospitalization has more than doubled since 1973. A conservative estimate for annual health care costs in this country would be $40–80 million. In addition, the Canadians estimate that 1 person in 1000 will spend 1 day per year in a hospital because of parasitic disease.[4] These physical, psychological, and health care costs are important to remember when comparing the amount of time, energy, and money spent for parasitic disease research.

In this paper we plan first to discuss the various helminthic infections of man in the United States followed by the protozoan infections. Then we will draw some conclusions about the importance of parasitic diseases in our society today.

Ascaris, Trichuris, and Hookworm

Ascaris, Trichuris, and hookworm, the most widespread helminth infections in the world, were extremely common in the rural areas of southern United States in the earlier part of this century. Since Warren's last review of the prevalence of these helminths nearly a decade ago, the situation has changed very little.[5] The few surveys that have been performed in the last decade reveal carriage rates for *Ascaris* and *Trichuris* only slightly less than those of 40 years ago (Table 3-1). Hookworm prevalence has declined substantially, probably as a result of health education, sanitation, and general development in the rural areas. The hookworm prevalence in Kentucky has declined from 37% in 1914 to 9% in the 1930s to 4% in 1963 to less than 1% in 1980.[5,6] Young school children in economically deprived rural areas have the highest prevalence of these endemically transmitted worms. The transmission rate for *Ascaris* remains great in the southern United States. Within 7 months following successful treatment of infection, more than two-thirds of the children were reinfected.[22] Unless specific control measures are undertaken, transmission will surely continue. *Ascaris* is the most potentially dangerous infection, occasionally causing bile duct or intestinal obstruction or eosinophilic pneumonitis.[7]

Based on a rural population of 27 million in the southern United States, it can be estimated that 2.5 million persons are endemically infected with *Ascaris* and an equal number with *Trichuris.* During the past decade approximately 3.5 million refugees and immigrants from Asia and Central and South America have entered the United States, contributing probably another 0.2 million individuals infected with *Ascaris, Trichuris,* or hookworm[23] (Table 3-1).

TABLE 3-1. Prevalence of helminths in studies published since 1972 (%)

	Ascaris	Trichuris	Hookworm	Strongyloides
A-US Schoolchildren				
Kentucky[6]	14	13	0.2	3
Louisiana[7]	2	NR	NR	NR
West Virginia[8]	16	3	0	0
Louisiana[9]	5	12.3	0.4	0.3
South Carolina[10]	16.5	14	0.1	0
South Carolina[11]	4	4	0	0
Louisiana[12]	21	56	0	0
B-Immigrant Groups				
Indochinese[13]	14	12	18	9
Puerto Ricans[14]	2	26	5	0
Southeast Asian refugees[15]	8	12	65	0
Indochinese[16]	13	9	7	1
Laotian refugees[17]	11	16	61	0
South & Central Americans[18]	3	32	5	0
Samoans & Filipinos[19]	9.5	30.0	4.3	0
South American[20]	0	14	6.6	1.7
Indochinese refugees[21]	20	13	17	7

*NR—not reported

Toxocariasis

Toxocara spp., the common roundworms of dogs and cats, are capable of infecting human beings and producing serious disease. Because of the popularity of pets in American society, there is widespread contamination of our environment with infective eggs. More than 99% of pups are infected due to the high efficiency of transplacental and transmammary transmission of *T. canis* larvae.[24] The prevalence of intestinal infection in adult pet dogs falls to less than 20%, but is higher among strays and unwanted pets. The prevalence of *T. cati* in cats varies from 24% to 67% and is highest among kittens.[24] Studies in various parts of the United States have demonstrated high rates of soil contamination in parks, playgrounds, and other public places.[24–26] Eggs may survive for years depending on soil type and climate.

Through intimate association with dogs and cats and ingesting eggs containing *Toxocara* spp., humans acquire this parasite and may develop visceral larva migrans, a syndrome characterized by fever, pulmonary symptoms, hepatomegaly, and eosinophilia. Occasionally, the larvae migrate to the eyes, producing granulomatous retinal lesions, and to the brain, producing encephalitis. More than 1900 cases of this disease have now been reported worldwide to date.[24] The actual prevalence of infection in the United States is difficult to determine as definitive diagnosis usually requires liver biopsy and serial sections. Serological surveys have used the ELISA test to yield crude prevalence estimates of between 0.5% and 3% of healthy persons. The largest survey of 8457 serum specimens from a representative sample of the U.S. population revealed a prevalence of infection of 3% using the ELISA. Even though this test has a remarkably good sensitivity and specificity of 78% and 92%, the proportion of screened persons with a positive test who actually have toxocariasis would be only 20% of this figure.[27] These results yield a prevalance of approximately 1,400,000 infected individuals in the United States today, but certainly a much smaller number suffer from severe disease.

Enterobiasis

Enterobius vermicularis, or the pinworm, is the most common helminthic infection in the United States. It is found in all socioeconomic groups and in all parts of the country.[5] Prevalence is greatest in children between 5 and 9, reaching extremely high levels in institutions, but enterobiasis is found in almost all age groups.[28] A conservative overall estimate based on lack of recent knowledge would be 20 million persons infected,[29] although Warren estimated 42 million about a decade ago.[5] A large proportion of cases is apparently asymptomatic, but some children suffer from anogenital pruritus and its concomitants.[30] Rare complications may occur. The major problem associated with this infection, however, is the psychological trauma, not only to the infected individual, but often to the entire family.

Strongyloidiasis

Strongyloides stercoralis, while not highly prevalent, is found throughout the country. It is probably most common in the rural South, infecting almost 3% of the population in some areas (Table 3-1)[6]. Strongyloidiasis may be found with particular frequency in institutions, e.g., in 18% of 1436 mentally retarded inmates in New York City.[31] The most common symptoms and signs in those with *Strongyloides* infection are abdominal pain, nausea and vomiting, weight loss, diarrhea, and urticaria.[32–35] Because of

its capacity to multiply in man via autoinfection, *Strongyloides* can paras-
itize man for more than 35 years.[32,34]

Many of the prisoners of war from the Far East in World War II and the
Vietnam conflict returned infected and will remain so until specifically
treated.[32,34,36] Because of autoinfection, *Strongyloides* is a potentially dangerous organism which may produce heavy infections leading to malabsorption, septicemia, and death. Recently a growing number of cases of
fatal, massive autoinfection have been reported in patients being treated
with corticosteroids, immunosuppressive therapy, or cytotoxic drugs.[33,35]
The number of subjects infected with *Strongyloides* has probably
remained constant over the last decade.

Trichinosis

Within the last decade there has been little change in the incidence of
trichinosis in humans. Between Stoll's estimated prevalence of 21.1 million for *Trichinella spiralis* infections in the United States and Canada in
1949[1] and Warren's estimated prevalence of 4.4 million infected persons
in 1972,[5] the incidence of reported cases of trichinosis dramatically
declined, paralleling the decline in prevalence of trichinosis found in
humans at autopsy—12–2.2%.[1,5,37,38–52] The decline in prevalence of trichinosis in humans paralleled a similar decrease (90%) in farm-raised hogs
and garbage-fed swine.[38] Since 1970, approximately 120 cases have been
reported yearly to the CDC. The actual number of infections occurring
yearly is probably 100–1000 times greater. Similarly, there are no indications that prevalence of the parasite in pork or humans has declined further than the estimates of 1972.[38] Pork continues to be the source in over
80% of the outbreaks, although ground beef intentionally or unintention-
ally adulterated with pork, bear, and walrus meat cause a small percentage
of cases yearly.[38] Alaska has the highest incidence per capita of any state
in the United States, usually from the ingestion of bear meat. Freezing
meat at 5° F for 20 days appears to kill the strains of *T. spiralis* from domestic swine, but larvae in infected Alaskan bear meat can be held for more
than 35 days without loss of infectivity.[38] It should be remembered that
trichinosis is a potentially lethal disease depending on the size of the
inoculum.

Tapeworms

With respect to the major tapeworm species infecting man, *Taenia saginata* is by far the most highly endemic giant cestode in the United States.
From 1967 to 1974, 0.05% of federally inspected cattle in the United States
had cysticercosis, while during the decade 1967–1977, the percent of
slaughtered beef cattle with cysts declined from 0.12 to 0.03. This means

approximately 15,000 infected carcasses are detected yearly and portions of the entire carcass are condemned.[4,39] Occasionally epizootics occur, such as the one reported from Arizona in 1975.[39]

T. saginata or *T. solium* ova or proglottids were identified in less than 0.1% of stool specimens submitted to State Health Laboratories for examination.[40] These specimens are from symptomatic individuals in whom the physician suspects parasitic infection. Most infections with this tapeworm are probably asymptomatic, but the number of individuals carrying this cestode has probably declined markedly in the last decade.

Taenia solium, the pork tapeworm, is rarely seen in the United States. Occasional cases of cysticercosis are seen in migrants from Central America. In Mexico it is one of the most common causes of space-occupying lesions in the brain.[41]

The fish tapeworm *Diphyllobothrium latum* has been indigenously transmitted in the Great Lakes but is rarely seen in men.

Hymenolepis nana, the dwarf tapeworm, is the most frequently identified tapeworm in stool specimens.[40] It occurs most frequently in children in the southern United States and in recent immigrants. Overall, the number of subjects infected with this helminth has probably remained about the same in the last decade.

Echinococcus

Man is the intermediate host for *Echinococcus granulosus,* and when infected harbors a slowly enlarging, space-occupying cyst. In Utah alone during the period 1944–1974, 56 indigenous cases of human disease occurred (33 between 1970 and 1974).[42,43] Almost an equal number was seen in Arizona, New Mexico, and California, where the parasite is endemic. During 1971–1974, 15.4% of dogs harbored tapeworms in three counties of Utah and 12% of 5631 sheep slaughtered at four abattoirs contained hydatid cysts. Incidence of hydatid disease in this part of Utah was comparable to that in other major endemic areas of the world.[42,43]

A sylvatic cycle of *E. granulosus* involving wolf, moose, caribou, reindeer, and incidentally, dogs, and Eskimos, has been described in Alaska. One hundred one human cases were reported from 1950 to 1966. Of the total aboriginal population of Alaska (50,000), it is estimated that 75% is at risk.[44]

Schistosomiasis

Although transmission of schistosomiasis cannot occur in the continental United States because of the lack of the snail intermediate host, Puerto Rican and Asian migration has brought in many cases. At the present time at least 1.5 million Puerto Ricans live in the continental United States, and approximately 10% of them carry schistosomiasis. Other migrant groups with a high prevalence include Yemeni, West Indians, South Americans, and Filipinos. Probably the total number of cases of schistosomiasis of 200,000 has not changed from 10 years ago.

Avian schistosomiasis is found in many areas of the United States. On penetrating the skin of man, the avian schistosome cercariae die. Repeated exposure may result in sensitization and the development of swimmers' itch, a pruritic papular rash. This problem has been reported from many parts of the United States including Oregon, Washington, Nebraska, and New York.[45]

Miscellaneous Helminth Infections

Examples of a wide variety of other worm infections may be easily found. Migrant Southeast Asians carry liver flukes.[21] The extremely common dog heart worm, *Dirofilaria immitis,* has caused over 50 reported cases of lung nodules, and related species such as *D. tenuis* are found in subcutaneous nodules.[46] Eosinophilic granulomas of the upper small intestine and stomach due to *Anisakis* ingested in raw fish (usually herring) are occasionally reported.[47] The state of Hawaii is in the area of the Pacific where the eosinophilic meningitis syndrome caused by the rat lungworm, *Angiostrongylus cantonensis,* is seen.[47] There are many other such worms sporadically infecting man which either become apparent because of the development of signs or symptoms of disease or are uncovered accidentally. The true prevalence of these infections in man in the United States remains unknown.

Toxoplasmosis

Infection with *Toxoplasma gondii* exists in chronic asymptomatic form in approximately one third of the population in the United States (Table 3-2). *T. gondii* is important for four major reasons: it may cause lymphadenopathy; as an opportunist, it may cause a lethal infection in the immunologically compromised host; it is responsible for at least 3000 congenitally infected infants in the United States yearly; and it is a major cause of chronic ocular inflammation.[48,49] The infection is a zoonosis, and members of the cat family are the definitive host. Man acquires the infection by ingesting either the cyst found in the uncooked meat of a variety of animals (Table 3-3) or by ingesting the mature oocysts excreted in the feces of an infected cat.[49,50,51] The organisms invade the intestinal epithelium and spread hematogenously to infect all organs, particularly the brain, heart, and skeletal muscle. Usually the infection is asymptomatic, but in the immunocompromised patient it is often fatal. Either the initial infection cannot be contained or there is reactivation of a latent infection resulting

TABLE 3-2. Prevalence of Sabin-Feldman dye test antibodies to toxoplasma in the United States population (% positive)

Age group (yr)	Portland OR	St. Louis MO	Syracuse NY	New Orleans LA	Pittsburgh PA	Military recruits
10–19	21	20	13	27	30	
20–29	25	29	35	35	24	
30–39	23	33	36	42	45	
40–49	15	40	36	36	68	
50+	16	39	44	45	67	
17–23						14

Modified from Feldman HA, Miller LT (1956) Serological study of toxoplasmosis prevalence. Am J Hyg 64:320–335; and Feldman HA (1965) A nationwide serum survey of United States military recruits. VI. Toxoplasma antibodies. Am J Epidem 81(3):385–391.

in encephalitis, myocarditis, pneumonitis, and death. Acute toxoplasmosis acquired by the mother during pregnancy (2–6 per 1000 pregnancies yearly in the United States) may infect the fetus and result in mental retardation, epilepsy, spasticity, paralysis, deafness, and chorioretinitis in the newborn.[48] Acute toxoplasmosis can be prevented by heating meat to 60° C or freezing to −20° C. Hands should be washed after touching

TABLE 3-3. Isolation of *Toxoplasma gondii* from muscle of domestic animals

Species	Country	Proportion	% Positive
Sheep	Australia	8/32	(25)
	New Zealand	3/5	(60)
	United States	8/86	(9.3)
	Germany	6/50	(12)
	Denmark	7/31	(23)
Cattle	Czechoslovakia	8/85	(9.4)
	New Zealand	0/80	(0)
	United States	0/352	(0)
	Germany	0/574	(0)
	Denmark	0/30	(0)
Swine	United States	12/50	(24)
	Italy	18/60	(30)
	Denmark	10/29	(35)
	Japan	28/191	(14)

uncooked meat. Cat feces should be avoided. Fruits and vegetables may be contaminated with oocysts and should be washed.[48]

Amebiasis

Entamoeba histolytica, the causal agent for amebiasis, produces dysentery, intestinal ulcerations, and perforations and hepatic and other extraintestinal abscesses. Many patients with *E. histolytica* infections have no symptoms. However the disease can produce a range of symptoms from mild illness with anorexia, malaise, weight loss, discomfort or cramping pain in abdomen, flatulence, rectal tenesmus and alternating episodes of diarrhea and constipation to severe invasive disease producing necrotic ulcerations throughout the colon, profuse bloody diarrhea, vomiting, fever, abdominal tenderness, and even colonic perforation and peritonitis.[52] Colonic perforation and rupture of hepatic abscesses are the most common causes of death. Even following treatment of amebic colitis, the patient may be left with strictures and postdysenteric colitis.[52]

Probably 4–5% of all Americans carry ameba in their intestines. Few population-based surveys have been done and a third of laboratories can not even correctly identify the presence of *E. histolytica* in stool specimens.[53] Serologic evidence for invasive amebiasis (elevated indirect hemagglutination titer) has been found in 2% of 6,195 randomly selected specimens throughout the United States, in 4% of healthy military recruits, and in 1% of 3,033 general hospital patients.[54-56] Half of these elevations are false positives, and the test remains positive 3–5 years following cure of the infection. In addition, for every case of invasive disease at least 10–20 others excrete cysts asymptomatically.[55,56] The population groups with the highest prevalence rates are recent immigrants and refugees from South and Central America and Southeast Asia,[13-21] residents of rural Southeastern United States[6-11] and mental institutions, and male homosexuals. More than one-third of sexually active homosexuals carry *E. histolytica* and probably transmit the ameba venereally.[57-61]

Giardiasis

Infection with *Giardia lamblia,* a flagellated protozoan of the small intestine, produces symptoms ranging from asymptomatic cyst passage to severe malabsorption. Prominent symptoms include diarrhea, abdominal cramps, fatigue, weight loss, flatulence, anorexia, and nausea that may continue as long as several months.[62]

Giardiasis is endemic in the United States and only recently has been recognized as an illness transmitted in several waterborne outbreaks.[63,64] In stools submitted to public health laboratories for examination, *Giardia* is the most frequently identified organism (3.9%).[41] In several stool surveys for parasites in healthy populations, it was demonstrated in 2–22% of individuals.[65–67] Detection rates vary because of the different groups surveyed, the known variable excretion of *Giardia* in stools, and the different techniques used for stool exams. A single stool exam can detect only 35–50% of the infections while six exams will detect 70–90% of them.[65]

Several groups within the United States have an especially high risk of infection. In the past decade more than 7000 waterborne cases have occurred in more than 20 areas primarily in mountainous parts of New England, the Pacific Northwest, and the Rocky Mountains.[63,64] Most result from the consumption of untreated surface water or chlorinated but unfiltered water from small municipal water systems or semipublic water systems in recreational areas. The water supplies become contaminated from both animal and human sources and the cysts remain viable in the usual chlorine concentrations in the purification systems. Rural children in the Southeast and Appalachia, the traditional foci for endemic intestinal parasite transmission, frequently carry this parasite.[8]

Approximately 20% of homosexual males are infected, and 8–30% of the Indochinese refugees who have recently arrived in the United States carry the agent.[13,21,58,66,67]

Pneumocystis Carinii

Every year several hundred persons in the United States develop pneumonia caused by *Pneumocystis carinii,* and up to one-third of these individuals die.[68] The illness rarely occurs in healthy persons; immunodeficiency is the outstanding predisposing factor. The organism seems to be ubiquitous. Sparsely scattered cysts may be found on occasion in hosts with no clinical or histological evidence of pneumonitis.[69] They appear to be in an inactive or latent state, but they may become activated if the host becomes immunosuppressed. In healthy children followed longitudinally, more than two-thirds acquire antibodies to the organism by 4 years of age.[70] Furthermore, a survey of more than 600 blood donors in the Netherlands demonstrated antibody titers in 41%.[69] Horizontal airborne transmission has been demonstrated in animal studies, but no definite confirmation has been made for man-to-man or animal-to-man transmission. The natural habitat for *P. carinii* remains unknown.[68]

The risk of developing clinical disease appears related to the extent of immunosuppression. The attack rates for children undergoing chemotherapy for cancer ranges from 5% for those receiving only one drug to 22% in the group receiving four anticancer drugs.[71,72] The risk is so high in those children receiving multiple anticancer drugs that trimethoprim-sulfamethoxazole is now routinely administered for chemoprophylaxis.[73]

In the past 4 years an epidemic of pneumonitis has occurred in seemingly healthy homosexual men, but when carefully tested they are all severely immunosuppressed for unknown reasons.[74-77] As noted, this population of homosexuals also has a very high incidence of other parasitic infections including amebiasis, giardiasis, and ascariasis.

Trichomoniasis

The prevalence of *Trichomonas vaginalis* has probably increased in recent years paralleling the epidemic incidence of other venereal infections. The infection rate tends to be higher in populations with greater numbers of sexual partners and is as high as 70% in prostitutes.[78] The overall proportion of women infected probably lies between 9% and 20%.[79-81] Between 30% and 80% of the male sexual partners of infected women carry the organism.[82] Only about one-half of the infected women have any symptoms, and infected men rarely if ever have any clinical manifestations.[83] The proportion of symptomatic individuals is probably higher than with *Toxoplasma* or *Entamoeba* infections, but in contrast, no one dies of trichomoniasis. The women may be acutely uncomfortable with dysuria, foul discharge, local vulvovaginal soreness, and irritation.[83] Fortunately treatment is fairly simple with a single dose of metronidazole.[84] We can estimate the number of infected women as 10% of those of childbearing age (14–44 years) and conservatively 30% of their male sex partners (5.2 million plus 1.6 million equals 6.8 million).

Malaria

In 1980, almost 2000 cases of malaria were reported to the Centers for Disease Control in the United States; this number represents a 250% increase from the number reported in 1979.[85] The majority of the cases occurred in refugees from Southeast Asia, but nevertheless more than 200 cases occurred in American civilians returning from abroad. Figure 3-1 illustrates the dramatic increase in incidence of malaria within the last decade; the 1980 reported figure is completely off the scale.[85] Each year several cases are acquired from blood transfusions, laboratory accidents, or congenital transmission. Endemic malaria transmission was terminated in the United States by 1950. Since then, only 13 isolated episodes of autochthonous malaria has occurred, despite the influx of thousands of infected Vietnam war veterans and immigrants from endemic areas.[86] Each year several people die or suffer severe cerebral, renal, or hemolytic complications from delay in recognition and treatment of the infection.

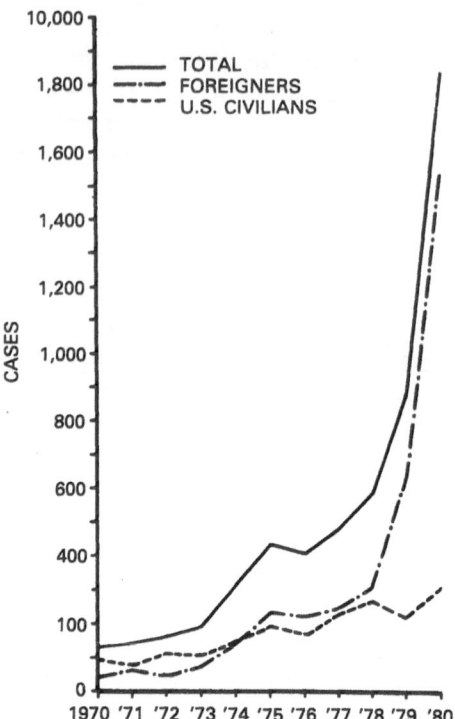

FIGURE 3-1. Cases of malaria in U.S. civilians and foreigners, United States, 1970–1980. (Courtesy of the Centers for Disease Control, Atlanta, Georgia.)

Miscellaneous Protozoan Infections

Finally, there are several protozoan infections which are seen relatively rarely in the United States. Fortunately, only a few cases of amebic meningoencephalitis caused by *Naegleria* or *Acanthamoeba* occur yearly since it is almost always fatal.[87] Leishmaniasis and trypanosomiasis are seen only in immigrants or travelers from abroad. Babesiasis, transmitted by ticks in the eastern United States, may be confused with malaria in its initial presentation but is usually self-limited except in asplenic individuals. *Balantidium coli* and *Dientamoeba fragilis* and *Cryptosporidium* can be found in a few individuals with persistent diarrhea.[87]

Conclusion

In conclusion, there is an enormous burden of worm and protozoan infection in the United States (Table 3-4). Moreover, it is doubtful, that there will be any striking changes in the near future because of the present health, economic and political priorities and the lack of highly efficient control methods. While many of the infections are benign, some cause sig-

TABLE 3-4. Prevalence of parasitic infections in 1947 (U.S. and Canada), in 1972, and in 1981 (U.S. alone)

	1947	1972	1982
Ascariasis	3.0	4.0	2.7
Trichuriasis	0.4	2.2	2.7
Hookworm	1.8	0.7	0.2
Enterobiasis	18.0	42.0	20.0
Strongyloidiasis	0.4	0.4	0.4
Trichinosis	21.0	4.4	4.4
Taeniasis (saginata)	0.1	0.2	<0.1
Hymenolopis	0.1	0.1	0.1
Schistosomiasis	—	0.2	0.2
Toxocariasis	—	—	1.4
Giardiasis	—	—	10.0
Amebiasis	—	—	10.0
Toxoplasmosis	—	—	70.0
Pneumocystis	—	—	154.0
Trichomoniasis	—	—	6.6

nificant morbidity and mortality. Enclaves of high prevalence and intensity of infection exist usually among the poor and the young, in minority groups, and in institutionalized individuals.

Under these circumstances, and without consideration for tourists, troops, and tropical populations, the high prevalence of worms in the American people cannot be ignored. Medical students should be made aware of the presence of these infectious agents and their unique characteristics. They should be taught how to manage helminthic and protozoan infections. Finally, more research by physicians and parasitologists is necessary to develop a better understanding of the host–parasite relationship and how to manipulate it to the point where these infections will largely disappear.

References

1. Stoll NR (1947) This wormy world. J Parasitol 33:1–18
2. Chernin E (1967) Epidemiology and prevention of selected parasitic diseases in the United States. In: Clark DW, MacMahon E (eds) Preventive medicine. Little, Brown and Co, Boston, pp 567–589
3. Neva FA (1972) Parasitic diseases of the GI tract in the United States. DM (Disease-a-Month) June: 3–44
4. Croll NA, Gyorkos TW (1979) Parasitic disease in humans: the extent in Canada. Can Med Assoc J 120: 310–312

5. Warren KS (1974) Helminthic diseases endemic in the United States. Am J Trop Med Hyg 23(4): 723–730

6. Walzer P, Milder JE, Banwell JG, et al (1982) Epidemiologic features of *Strongyloides stericoralis* infection in an endemic area of the United States. Am J Trop Med Hyg 31(2): 313–319

7. Blumenthal DS, Schultz MG (1975) Incidence of intestinal obstruction in children infected with *Ascaris lumbricoids*. Am J Trop Med Hyg 24(5): 801–805

8. Walker RB, Hough JC, Brough JW (1980) A survey of intestinal parasites in rural children. J Fam Pract 11(4): 559–561

9. Morgan PM, et al (1972) Intestinal parasitism and nutritional status in Louisiana. J La Med Soc 124(6): 197–203

10. Jacobs CF, Teator DMN, Jacobs NE (1972) A survey of intestinal parasites in Charleston County school children. J SC Med Assoc 68(8): 315–319

11. Penick RW, et al (1976) Pinworm infestation: a statewide study. J of the SC Med Assoc 72(9): 348–349

12. Blumenthal DS, et al (1976) Effects of *Ascaris* infection on nutritional status in children. Am J Trop Med Hyg 25(5): 682–690

13. Hoffman SF, Barrett-Connor E, Norcross W, Nguyen D (1981) Intestinal parasites in Indo-Chinese immigrants. Am J Trop Med Hyg 30(2): 340–343

14. Ortiz JS (1980) The prevalence of intestinal parasites in Puerto Rican farm workers in Western Massachusetts. Am J Public Health 70(10): 1103–1105

15. Wiesenthal AM, Nickels MK, Hashimoto KG, Endo T, Ehrhard KG (1980) Intestinal parasites in S.E. Asian refugees. Prevalence in a community of Laotians. JAMA 244(22): 2543–2544

16. Boudreau E, Doberstyn EB, Colon A, Tina L, et al (1980) Health screening of resettled Indo-Chinese refugees. Washington, D.C. Utah MMWR 29(1): 4, 9–11

17. Lindes C (1979) Intestinal parasites in Laotian refugees. J Fam Pract 9(5): 819–822

18. Hargus, EP, Lepow M, Lau T, Colon AR (1976) Intestinal parasitoses in childhood populations of Latin origin. Lessons from a survey of 129 such children in Hartford, CT. Clin Pediatr (Phila) 15(10): 927–929

19. Desowitz RS, Wiebenga NH (1975) A survey of intestinal parasites in Oahu school children. Hawaii Med J 34: 21–23

20. Winsberg GR, Sonnenschein E, Dyer AR, Schnadig V, et al (1975) Prevalence of intestinal parasites in Latino residents of Chicago. Am J Epidem 102(6): 526–532

21. Parasitic infections in Indo-Chinese refugees, Pinellas County, Florida. Public Health Service-CDC-Atlanta. EPI-79-92-2. June 3, 1981

22. Miller MJ, et al (1978) An evaluation of levamisole for treatment of ascariasis. South Med J 71(2): 137–140

23. Statistical Abstracts of the United States, 1980. 101st ed. US Dept of Commerce. Bureau of Census

24. Glickman LT, Schantz PM (1981) Epidemiology and pathogenesis of zoonotic toxocariasis. Epidemiologic Reviews 3: 230–250

25. Dubin S, Segall S, Martindale J (1975) Contamination of soil in two city parks with canine nematode ova including *Toxocara canis:* a preliminary study. Am J Public Health 65: 1242

26. Surgan MH, Colgan KB, Kennett SI, et al (1980) A survey of canine toxocariasis

and toxocaral soil contamination in Essex County, New Jersey. Am J Public Health 70: 1207

27. Glickman L, Schantz P, Dombroski R, et al (1978) Evaluation of serodiagnostic tests for visceral larva migrans. Am J Trop Med Hyg 27: 492–498

28. Amin OM, et al (1980) Prevalence of pinworm and whipworm infestations in institutionalized mental patients in Wisconsin, 1966–1976. Wis Med J 79(3): 31–32

29. Return of the pinworm. Med Times 103(2): 178

30. Wolfe MS (1978) Oxyuriasis, Trichostrongylus and Trichuris. In: Marsden P (ed) Clinics in Gastroenterology 7(1): 201–219

31. Yoeli M, Most H, Berman H, Tesse B (1963) The problem of strongyloidiasis among the mentally retarded in institutions. Trans R Soc Trop Med Hyg 57: 336–345

32. Gill GV, Bell DR (1979) *Strongyloides stercoralis* infection in former Far East prisoners of war. Br Med J 2(6190): 572–574

33. Filho EC (1978) Strongyloidiasis. In: Marsden P (ed) Intestinal parasites. Clinics in Gastroenterology 7(1): 179–201

34. Grove DI (1981) Strongyloidiasis in allied ex-prisoners of war in S.E. Asia. Brit Med J 2(6214): 59801

35. Milder JE, Walzer PD, Killgore G, Rutherford I, Klein M (1981) Clinical features of *S. stercoralis* infections in an endemic area of the United States. Gastroenteralogy 80: 1481–1488

36. Berg SW, Richlin M (1977) Injuries and illnesses of Vietnam war POWs. II. Army POWs. Milit Med 142(8): 598–602

37. Zimmerman WJ (1974) The current status of trichinellosis in the United States. In: Kim CW (ed) Trichinellosis. Intext Educational Publishers, New York, pp 603–609

38. Centers for Disease Control (1981) Trichinosis surveillance. Annual summary—1980

39. Slonka GF, et al (1975) An epizootic of bovine cysticercosis. J Am Vet Med Assoc 166(7): 678–681

40. Center for Disease Control (1978) Intestinal Parasite Surveillance. Annual Summary

41. Jones TC (1978) Cestodes. Clin Gastroenterol 7(1): 105–128

42. Pappaioano M, Schweke CW, Sard DM (1977) An evolving pattern of human hydatid disease transmission in the U.S. Am J Trop Med Hyg 26(4): 732–742

43. Schantz PM (1977) Echinococcus in American Indians living in Arizona and New Mexico. Am J Epidemiol 106(5): 370–379

44. Wilson JF, Diddams AC, Rausch RL (1968) Cystic hydatid disease in Alaska. A review of 101 autochthonous cases of *E. Granulosus* infections. Am Rev Resp Dis 98: 1–15

45. Cort WW (1950) Studies on schistosome dermatitis. XI. Status of knowledge after more than twenty years. Am J Hyg 52: 251–307

46. Ciferri F (1981) Human pulmonary dirofilariasis in the West. West J Med 134(2): 158–162

47. Marsden P (1978) Other nematodes. Clin Gastroenterol 7(1): 219–229

48. Krick JA, Remington JS (1978) Current concepts in parasitology. Toxoplasmosis in the adult—an overview. N Engl J Med 298(10): 550–553

49. Remington JS, Desmots G (1976) Toxoplasmosis. In: Remington JS, Klein JO

(eds) Infectious diseases of the fetus and newborn infant. WB Saunders, Philadelphia, pp 191–332

50. Teutsch SM, Juranck DD, Sulzer A, Dukey JP, Sikes RK (1979) Epidemic toxoplasmosis associated with infected cats. N Engl J Med 300(13): 695–699

51. Dubey JP and Streitel RH (1976) Prevalence of *Toxoplasma* infection in cattle slaughtered in an Ohio abattoir. J Am Vet Med Assoc 169(11): 1197–9

52. Juniper K (1978) Amoebiasis. Clin Gastroenterol 7(1): 3–30

53. Ruebush TK, Juranek DD, Brodsky RE (1978) Diagnoses of intestinal parasites by state and territorial public health laboratories, 1976. J Infec Dis 138(1): 114–117

54. Sargeaunt PG, Williams JF (1982) A study of intestinal protozoa including nonpathogenic *Entameba histolytica* from patients in a group of mental hospitals. Am J Public Health 72(2): 178–180

55. Healy GR (1976) The status of invasive amebiasis in the United States as determined by studies with the indirect hemaglutination test. In: Sepulveda B, Diamond LS (eds) Amebiasis. Instituto Mexicano del Seguro Social, Mexico, pp 619–628

56. Kagan IG (1976) Seroepidemiology of amebiasis. In: Sepulveda B, Diamond LS (eds) Amebiasis. Instituto Mexicano del Seguro Social, Mexico, pp 574–588

57. Marr JS (1981) Amebiasis in New York City: a changing pattern of transmission. Bull NY Acad Med 57: 187–200

58. Keystone JS, Keystone DL, Proctor EM (1980) Intestinal parasitic infections in homosexual men: prevalence, symptoms and factors in transmission. Can Med Assoc J 123: 512–514

59. Quinn TC, Corey L, Chaffie RG, Schuffler MD, et al (1981) The etiology of anorectal infections in homosexual men. Am J Med 71: 395–406

60. Pomerantz BM, Marr JS, Goldman WD (1980) Amebiasis in New York City 1958–1978. Bull NY Acad Sci 56(2): 232–244

61. Identification of the male homosexual high risk population. (1981) Bull NY Acad Med 56(2): 232–243

62. Smith JW, Wolfe MS, (1980) Giardiasis. Ann Rev Med 31: 373–383

63. Juranck D (1979) Waterborne Giardiasis. In: Jakubowski W, Hoff JC (eds) Symposium on waterborne transmission of giardiasis. Environmental Protection Agency, Cincinnati, pp 150–163

64. Croun GF (1979) Waterborne outbreaks of giardiasis. In: Jakubowski W, and Hoff JC (eds) Symposium on waterborne transmission of giardiasis. Environmental Protection Agency, Cincinnati, pp 127–149

65. Healy GR (1979) The presence and absence of *Giardia lamblia* in studies on parasite prevalence in the USA. In: Jakubowski W, and Hoff JC, (eds) Symposium on waterborne transmission of giardiasis. Environmental Protection Agency, Cincinnati, pp 92–103

66. Kean BH, William DC, Luminais SK (1979) Epidemic of amoebiasis and giardiasis in a biased population. Br J Vener Dis 55: 375–378

67. Phillips SC, Mildran, P, Williams DC, Gelb AM, White MC (1981) Sexual transmission of enteric protozoa and helminths in a venereal-disease-clinic population. N Engl J Med 305: 603–606

68. Hughes WT (1977) *Pneumocystis carinii pneumonia.* N Engl J Med 297(25): 1381–1384

69. Ruskin J (1976) *Pneumocystic carinii.* In: Remington JS, and Klein JO (eds)

Infectious diseases of the fetus and newborn infant. WB Saunders, Philadelphia, pp 691–746

70. Pifer LL, Hughes WT, Stagno S, Woods D (1978) *Pneumocystis carinii* pneumonitis infection: evidence for high prevalence in normal and immunosuppressed children. Pediatrics 61(1): 35–41

71. Hughes WT, Price RA, Kim HK, Coburn TP, et al (1973) *Pneumocystic carinii* pneumonitis in children with malignancies. J Pediatr 82(3): 404–415

72. Hughes WT, Feldman S, Aur RJA, et al (1975) Intensity of immunosuppressive therapy and the incidence of *Pneumocystis carinii* pneumonitis. Cancer 36: 2004–2007

73. Hughes WT, Kuhn S, Chaudhary S, Feldman S, et al (1979) Successful chemoprophylaxis for *Pneumocystis carinii* pneumonitis. N Engl J Med 297(26): 1419–1426

74. Center for Disease Control (1981) Kaposi's sarcoma and *Pneumocystis* pneumonia among homosexual men—New York City and California. MMWR 30: 305–308

75. Center for Disease Control. Task Force on Kaposi's Sarcoma and Opportunistic Infections (1982) Epidemiologic aspects of the current outbreak of Kaposi's sarcoma and opportunistic infections N Engl J Med 306: 248–252

76. Gottleib M, Schroff R, Schanker H, et al (1981) *Pneumocystis carinii* pneumonia and mucosal candidiasis in previously healthy homosexual men. N Engl J Med 305: 1425–1431

77. Masur H, Michelis NA, Greene JB, et al (1981) An outbreak of community-acquired *Pneumocystis carinii* pneumonia: Initial manifestation of cellular immune dysfunction. N Engl J Med 305: 1431–1438

78. Gallai Z, Sylvestre L (1976) The present status of urogenital trichomoniasis: a general review of the literature. Appl. Therapeutics 773

79. Schnell JD (1974) The incidence of vaginal candida and trichomonas infections and treatment of trichimonas vaginitis with clotrimazole. Postgrad Med J 50(Suppl 1): 79–81

80. Eriksson G. et al (1975) Frequency of *N. gonorrheae, I, vaginalis* and *C. albicans* in female venerealogical patients. A one-year study. Br J Vener Dis 51(3): 192

81. Lazar A (1970) *Trichomonas vaginalis* infection. Incidence with use of various contraceptive methods. J Med Soc NJ 67: 225–226

82. Weston TET, Nicol CS (1963) Natural history of trichomonal infections in males. Br J Vener Dis 39: 251–257

83. Fouts AC, Kraus SJ (1980) *Trichomonas vaginalis:* Reevaluation of its clinical presentation and laboratory diagnosis. J Infect Dis 141(2) 137–143

84. Dykers JR (1975) Single dose metronidozale for *T. vaginitis.* N Engl J Med 293: 23–24

85. Centers for Disease Control. (1980) US Department of Health and Human Services. Public Health Service. Malaria Surveillance. Annual Summary 1979.

86. Taylor AF, Gaspers S, Mahoney LE, et al (1982) Introduced autochthonous *Vivex* malaria—California, 1980–1981. MMWR 31(16): 213–215

87. Jones TC (1979) Freeliving amoeba. In: Mandell G et al. (eds) Infectious diseases. John Wiley, New York, pp 2095–2097

4

The Importance
of Parasitic Diseases

Andrew Davis

Within the broad context of this volume, I have chosen to write on the importance of parasitic diseases for three reasons:

1. they have been with us since time immemorial;
2. they will be with us for an unpredictable length of time and certainly after the year 2000, which WHO has chosen for its slogan "Health for All";
3. anyone even remotely familiar with the work of WHO will be aware of the Organization's interest and activities in parasitic diseases in the last 34 years, since its foundation.

So my paper will be targeted on the current status of parasitic diseases. Throughout the developing countries in both tropical and subtropical climates, parasitic diseases remain among the most ubiquitous and serious public health problems with depressingly high prevalence rates of the major protozoal and helminthic infections.

Parasitic infections may produce a broad spectrum of diseases of major pathogenic or social importance. Some protozoal infections such as African trypanosomiasis are fatal if untreated, while others, such as amoebiasis on a global scale, or Chagas' disease in South America, may produce significant acute or chronic disease in a high proportion of young populations. Helminth infections tend to produce episodic periods of diagnosable ill health against a background of slow cumulative morbidity, and chronic pathological sequelae of schistosomiasis and filariasis are well recognised throughout endemic areas.

What is perhaps not so well appreciated is the impact of these diseases on the economy of developing countries by the production of high morbidity rates, lost working days, decline in output at both individual and community level, and nonproductive allocation of sparse financial resources for curative treatment, which is frequently nullified because the prevailing environmental and socioeconomic background conditions lead to reinfection and perpetuation of epidemiological transmission cycles.

Global estimates of prevalences of parasitic diseases are of such staggering magnitude that the mind has difficulty in the conception of the descriptive statistics; the total number of protozoal and helminthic infections currently existing far outnumbers the world population since multiple infections are the rule, not the exception, in most tropical areas. Statistical estimates of infection rates merely represent the rates of diagnosis, which depend on the sensitivity of the techniques used. Hence the true prevalence of the parasitic diseases may be much higher than the official estimates, and this fact, combined with the known inadequacy of health statistics over much of the world, makes it highly probable that the parasitic diseases have, in numerical terms, been grossly underestimated in the past.

In reviewing the major parasitic diseases, the official estimates make chilling reading. The Sixth Report on the World Health Situation 1973–1977[1] notes that malaria, the greatest killer among the tropical diseases, is also the most widespread and is a major public health problem in some 70 countries around the world. Where epidemics occur, malaria is also of socioeconomic and political importance. Each year in Africa at least a million children under 14 years of age die from malaria complicated by nutritional and other health problems. The impact of available control measures can be judged by extrapolation of the results of a controlled study of insecticidal spraying operations in Kenya, when general mortality decreased in the two years from 1973 to 1975 from 23.9 to 13.5 deaths per 1000 population and infant mortality from 157 to 93 deaths per 1000 live births.[2] Yet, despite these encouraging figures, the wide variety of ecological situations in the world nullifies one overall approach to control. A specific epidemiological analysis must be conducted in each situation in order to decide on the optimal tactics to be adopted. For example, in contrast to the Kenya study is the situation in the Sudan savanna areas of Africa, where malaria is pandemic. Each inhabitant of these areas receives between 40 and 120 infective mosquito bites per year. It is unlikely that the currently available drugs and insecticides would interrupt transmission, nor can the endemic countries afford the expenses involved. Hence a tactical variant of morbidity control by chemotherapy to the population groups most at risk would be adopted.

Of a total of 2,015 million people living in areas where malaria has been or is still endemic, some 824 million are in locations where malaria has been eradicated, 848 million are in areas where eradication or control programmes are currently in progress, and 343 million are in endemic areas unprotected by any specific antimalaria measures, the majority in Africa.

Despite these impressive figures brought about by the immense efforts of the WHO malaria eradication campaigns of the 1950s and 1960s, there has been, in recent years, a serious resurgence of malaria in Southeast Asia and Latin America.

Moving on to the other major parasitic diseases, I regret to say that the picture, in human terms, remains gloomy. Epidemiologically many of the parasitic infections are advancing, not receding, and the paradox is that this is occurring despite the technological revolution of the 20th century. Parasitic diseases have been becoming increasingly recognized as complex syndromes, often not amenable to simple control measures because they are intimately associated with human behavioural factors, human agricultural practices, man's domestic and work animals, and the more general fields of environment, sanitation, education, poverty, and socioeconomic structures. Many have complicated transmission cycles, including some with reservoirs in the animal world.

The combination of the relentless population increase in the developing world and the necessity for increased food production dependent on agricultural practices relying on irrigation has increased the prevalence, incidence, and intensity of schistosomiasis. The construction of man-made lakes for agricultural development or expansion and the production of the hydroelectric power so essential for economic advancement has aggravated these factors and will continue to do so.

Urbanisation, with rural populations attracted to towns in search of work, inevitably results in an outstripping of the available sanitary facilities, often inducing culicine breeding, the production of new foci of transmission of filariasis, and extension of preexisting gastrointestinal protozoal and helminthic infections.

Political upheavals, wars, and population disruptions in Africa all destroy well-established surveillance systems and are known to cause epidemics of trypanosomiasis. Paradoxically, clearance of scrub or bush land for settlements and development may disturb enzootic foci of leishmaniasis and transmission to man may result.

Above all, man himself, through constant and increasing population movement, may create new foci or intensify preexisting foci of parasitism in areas where vectors or intermediate hosts exist, as is the case in the vast majority of the tropical areas.

Schistosomiasis

Schistosomiasis is essentially an infection of areas, usually rural and agricultural but also peri-urban in many tropical countries, where there exists poverty, ignorance, poor housing, substandard hygienic practices, and few, if any, sanitary facilities. Local agricultural techniques may be based on the use of natural fresh waters or primitive irrigation systems which offer fertile sites of transmission. The disease becomes an occupational hazard for peasant farmers. Children act as important human reservoirs of infection by urinating indiscriminately in or near fresh water, while faecal pollution of water as a result of inadequate excretal disposal facilities is the most important factor in maintaining transmission of the intestinal types of schistosomiasis. In developing countries, activities other than occupational also bring the population into contact with water, where transmission begins; examples can be found easily in domestic, hygienic, agricultural, recreational, and religious customs.

Possibly 200 million people are infected and 500–600 million exposed to the threat of infection. The geographic distribution is wide, ranging from China, the Philippines and Indonesia *(Schistosoma japonicum)*, the Arabian Peninsula and nearby states, the Sudan and the Nile Valley and Delta, numerous countries of the north African littoral, the whole of sub-Saharan Africa (*S. mansoni* and *S. haematobium*) to the New World (Brazil, Surinam, Venezuela, and certain Caribbean islands; *S. mansoni*).

The low direct mortality is offset by a very high prevalence of infection and by the multiplicity of chronic pathological sequelae of infection, which may underlie causes of death ascribed to other conditions. Hence morbidity due to schistosomiasis represents, in numerical terms, a profound burden on the public health and therapeutic services in countries where transmission is endemic.

The major sites of serious infections are in Brazil, China, Egypt, Sudan, and various countries of Africa (Ghana, Nigeria, Zaire, Morocco).

The striking advances in chemotherapy, particularly in the development of effective short-term drug regimens, of the last 15 years have induced a reappraisal of control strategy and tactics. Control measures currently available are effective and, where governments have recognised schistosomiasis as a priority health problem, where motivation of both government and population coincided, and where skilled and interested personnel were available, some striking control successes have been recorded.

Numerous other snail-borne trematode infections with complex transmission cycles exist. Some are major public health problems in Southeast Asia, China, and other countries in the Western Pacific region, where infection is acquired by age-old eating customs.

Filariasis

Lymphatic Filariasis

Over 250 million people are infected with one or more of the lymphatic-dwelling filarial parasites *Wuchereria bancrofti, Brugia malayi,* and *B. Timori.* Most suffer attacks of adenolymphangitis and filarial fever; a proportion develop elephantiasis or hydrocoele. Bancroftian filariasis is globally widespread and is a particularly severe problem in most tropical countries in southeastern Asia (particularly India), Africa, America, and the Pacific Islands.

Brugian filariasis is localised to southeastern Asia and is numerically most important in Indonesia. It is a more virulent parasite than *W. bancrofti* and is often difficult to control because animal reservoirs exist.

Control of both these forms of filariasis is possible using mass chemotherapy with diethylcarbamazine (DEC) but may be difficult to implement because of the reactions to treatment experienced by the patients. Ancillary measures to reduce sources of vector mosquitoes vary in feasibility with the genera involved (*Culex, Anopheles, Mansonia, Aedes* spp). Health education and measures to prevent man–mosquito contact can also be very effective.

Onchocerciasis

Between 20 and 40 million people are infected with *Onchocerca volvulus*. In different ecological zones varying proportions are at risk of blindness. In areas of intense infection, up to 15% of the population may be blind, and so the disease has severe socioeconomic effects. Many more suffer from severe itching and disfiguring skin lesions. The worst affected areas lie in tropical Africa, particularly in the sub-Saharan savanna belt, and in localised areas of Guatemala and Mexico. A mild form of the disease threatens to spread in tropical South America; a focus of severe skin and lymphatic disease occurs in Yemen. Currently control depends on larviciding against the vector *Simulium* spp. since chemotherapy with existing drugs is unsatisfactory.

Guinea Worm *(Dracunculus medinensis)*

This temporarily incapacitating condition, affecting agricultural workers particularly, still occurs in parts of Africa and Asia. Its control depends on the establishment of hygienic wells and water supplies.

Trypanosomiasis

African Trypanosomiasis

With some 10,000 new infections now reported annually, sleeping sickness may give the impression of being a health problem of only moderate importance. Its relatively low current incidence, however, is the result of persistent preventive measures implemented over the last 50 years, which consisted of regular medical surveillance and treatment, with or without tsetse fly control. These operations continue to receive high priority because of the severe nature of the disease: dramatic mental deterioration, 100% mortality if untreated, and the occurrence during treatment of side effects with a fatality rate of 5–10%.

Recent estimates place the number of people actually exposed to sleeping sickness risk at 45 million, of which only 6 million are under regular surveillance. Where regular surveillance is discontinued, patients will remain undetected until the central nervous system is severely infected, and as a result, man-to-man transmission increases and epidemics occur. A serious example is the severe outbreak in Uganda with over 10,000 new infections during 1980. This epidemic is still not under control.

Most of the sleeping sickness surveillance and vector control operations are performed by national health services which, despite many pressing health problems, continue to give high priority to trypanosomiasis control. They succeed in maintaining regular surveillance of 6 million people in the face of high costs (estimated at a total yearly investment of U.S. $5 million) and despite the permanent deployment of large numbers of health personnel and vehicles. The overall situation, however, gives rise to considerable concern, since the exposed population at risk without appropriate surveillance procedures has increased in recent years.

Chagas' Disease

In South America Chagas' disease is associated with traditional rural housing and poor sanitary conditions, which create appropriate habitats for the reduviids which transmit infections with *Trypanosoma cruzi.*

At present, some 50 million or more of the rural population in South America live under conditions favouring transmission. Accurate prevalence figures are scarce, and in some endemic countries, absent.

The proportion of infection in endemic areas can reach 50–70%, and the total number of infected people in the continent is at least 10 million. These figures are both minimal and far from accurate, due to the lack of systematic surveillance, underreporting, and inadequate diagnostic facilities.

Infection with *T. cruzi* is not necessarily followed by the development of manifest lesions, and the finding of positive serology, commonly used as a diagnostic criterion, is, therefore, of little prognostic significance. Initial infections are usually acquired in childhood, may be fatal in 5–10% of cases, and are frequently not identified as trypanosomiasis. A chronic phase may follow, up to 10–20 years later, with cardiac and/or intestinal lesions, leading to chronic disablement and eventually death. Since the recently developed chemotherapeutic agents against Chagas' disease appear to be active only in the early stage of infection, the treatment of the late stage, when the disease is commonly diagnosed, is generally unsuccessful.

Prevention could best be achieved by replacing traditional houses and by changing part of the living customs of the populations in endemic areas. Such an ultimate aim, however, reaches far beyond the immediate objective of a public health programme and, because of its social and economic implications, can only be envisaged as a solution for the distant future.

As interim measures, prevention of the disease by regular house-to-house spraying of insecticides is practised on a large scale in Venezuela and Brazil. These methods are laborious and costly, however, and in Venezuela resistance of the transmitting bugs has been reported to occur. In most of the endemic areas, intervention measures are not implemented, and the majority of patients are hospitalized for symptomatic treatment in the advanced stage of the disease.

Recent cause for concern is the apparent spread of Chagas' disease to urban situations by migration of rural people to the periphery of the cities, and through transmission by blood transfusions from infected donors.

Leishmaniases

The leishmaniases cover a wide range of diseases broadly divided into visceral, mucocutaneous, and cutaneous leishmaniases. Transmission is mainly through the bite of various species of sandflies. The group of diseases includes anthroponotic as well as zoonotic forms.

Among the various clinical syndromes produced, visceral leishmaniasis, or kala-azar, is a systemic infection with high mortality if untreated. The mucocutaneous form (espundia) causes serious lesions of the nasopharynx with extensive disfigurement, damage to vocal cords, and respiratory dysfunction. The cutaneous forms are the least serious group of the diseases, but, whilst usually self-healing, they can cause lifelong disfiguring scars. A rarer diffuse form of cutaneous leishmaniasis is a severe chronic and incurable disease, causing extreme disfiguration.

The distribution of the leishmaniases is worldwide within the tropics and subtropics, the total number of patients being unknown. According to recent conservative estimates, the present incidence of new infections may be of the order of 400,000 per year.

The public health importance of these diseases has been reemphasized in recent years. Outbreaks of visceral leishmaniasis in India and Kenya have shown the severity of the disease in terms of morbidity and mortality and have illustrated the difficulties in control. As a side benefit of malaria-control campaigns, anthroponotic leishmaniases practically disappeared, but as a result of the interruption of vector control operations in some areas, there has been a drastic resurgence in the leishmaniases. In addition, the diseases are apparently spreading, notably in East Africa.

The principal problems encountered in control are the high costs of insecticides, the lack of simple diagnostic procedures in kala-azar, and the increasing resistance of the parasites to currently available chemotherapeutic compounds.

Gastrointestinal Protozoal
and Helminthic Infections

Gastrointestinal parasitic infections are caused by a great number of protozoa and helminths that are either specific parasites of man, or zoonotic, or affecting both man and animals.

The most important parasites are the protozoa *Entamoeba histolytica* and *Giardia lamblia* and the nematodes *Ascaris lumbricoides, Trichuris trichiura,* and *Strongyloides stercoralis.* In addition to these practically cosmopolitan species of intestinal helminths, there are others of more local importance like *Taenia saginata, T. solium, Hymenolepis nana,* and *Diphyllobothrium latum.*

One-third of the work population is infected by *Ascaris* and one-quarter by hookworms. Many millions of people in Africa, Asia, and Latin America, especially those in rural areas, have multiple helminthic infections. Parasitism is particularly frequent and serious in undernourished populations, where it causes considerable damage through competitive utilization for scarce food material, intestinal malabsorption, and blood loss and aggravates caloric, protein, and vitamin deficiencies.

Amoebiasis also represents a worldwide problem, but the epidemiological picture is heterogeneous. In some countries it seems to occur only in healthy carriers and in the form of a fairly benign illness; in others amoebic dysentery and liver abscess are frequent. The existence of *E. histolytica*-like species with different pathogenic potentials makes etiological diagnosis particularly difficult and calls for new research. Some free-living species can facultatively infect man and rarely, fatal cases of meningoencephalitis are caused by the amoebae of the *Limax* group *(Naegleria; Hartmanella; Acanthamoeba).*

Having taken a bird's-eye view of some, and I stress the word some, of the tropical parasitic infections, we must ask ourselves what is their importance? This question immediately plunges us into a series of areas of greater of lesser interest, depending upon one's professional background, training, experiences, instincts, and skills.

The importance of parasitic diseases is widespread—in a medical context, in an epidemiologic context, in a human context, in a humanitarian context, in ecological and zoological contexts, in the veterinary context, in the agricultural context, in developmental and socioeconomic contexts, in the research context—the list appears endless.

As scientists, physicians, zoologists, entomologists, we should be stimulated by these considerations to stimulate the various educational, political, financial, and administrative authorities to enlarge the teaching, the study, and the support of research in parasitology, which is in danger of becoming a neglected field familiar only to devotees of the esoteric arts.

Perhaps physicians concerned with medical parasitology should learn from the veterinary and farming professions, where there is patently much less doubt and much more positive interest in diseases caused by parasites, possibly because the ill effects are more easily quantified and measured. While we do not believe in the absolute truth of statistics, we all must be impressed by the fact that, measured in kilogram of products per head of livestock population, the beef/veal table descends from a high of 93 for North America, 36 Latin America, 17 in the Near East, 15 in Africa, to 4 in the Far East. Or conversely, the regional consumption of animal protein in grams per head per day rises from 9.1 in the Far East, through 10.8 Africa, 13.4 Near East, 23.4 Latin America, 41.6 Europe, to 69 in North America. The paradox here is that the developing countries possess about 65% of the world's domestic animal resources but produce only some 30% of the world's meat.[3] The last 20 years has seen virtually a zero growth rate in the production of beef per head of animal in the developing countries. The increases in total animal protein production were due to increases in animal numbers rather than increased productivity per animal, and the scope for a continued increase in livestock numbers for ruminants is limited by the population explosion, unsuitable land, water shortage, and parasitic infections, the most striking of which is animal trypanosomiasis. Although it can be said that trypanosomiasis is probably the only disease which has profoundly affected the settlement and economic development of a major part of a continent,[3] it should also be appreciated that in the Third World, losses in poultry and livestock from internal and external parasites are very large. In some countries, calf mortality may be 30% or more of the annual calf crop; much attributable to parasitism, to which can be added the invisible losses such as the amounts of meat and milk not produced, the reduced length of productive life, and the loss of animal working capacity.[3]

Yet these constraints applicable to the Third World can be brought much nearer home by considering the helminthic infections well known, not only in the tropics, but throughout Europe, e.g., fascioliasis, bovine and porcine cysticercosis, or echinococcosis. There is little need to emphasize the causal roles of these parasites in producing economic loss to the livestock producer and the meat industry, quite apart from their importance in human diseases, where we still have no satisfactory therapy for hydatid disease or for many patients with neurocysticerosis.

I have restricted my remarks to those aspects of parasitology with which I can claim a limited familiarity, but outside of these areas, there are enormous fields of importance where parasitology and parasitic diseases play major roles.

Agriculture and human welfare are entwined. The availability of land for arable food production is frequently limited by the presence of endemic parasites—of plants as well as man. Third World ecological conditions frequently favour agricultural pests, and one author made a "guesstimate" that 48% of the world's agricultural production is lost to pests and diseases at pre- and post-harvest times.[4] Certainly plant nematode parasites are of great relevance in agriculture as a cause of crop damage and well known examples are the banana root nematode *(Radopholus sinilis), Meloidogyne javanica* and *M. incognita* affecting tobacco—on reflection perhaps no bad thing in human health terms—and pineapple, coffee, and coconut losses may be severe problems in struggling economies.

This brings us to socioeconomic complications of parasites and parasitism. Here the literature is voluminous, the measurements numerous, and the estimates of the ill-effects and the benefits of control as widely variable as one might anticipate. Here, for one brief period, let us heretically throw away science and concentrate on a few fundamentals, largely from the medical parasitology viewpoint.

The populations of the developing countries suffer from a multitude of disadvantages: poverty, malnutrition, low income, high birth rates, illiteracy, and parasitic diseases. Ill health results from, and causes, poverty, and all experience indicates that improvement in one set of causes of poverty, that is, ill-health, has positive effects on poverty itself.

Many developing countries are predominantly rural societies and have a heavy dependence on export earnings from agriculture. It is easy to see how tropical parasitic diseases undermine socioeconomic potential; examples are legion: malaria in India, filariasis in Sri Lanka, schistosomiasis in Sudan and Egypt, onchocerciasis in West Africa.

Control of parasitic diseases in the Third World, implemented on a decade-long International Programme of Water Supply and Sanitation currently launched, would have immense benefits to the Third World, to the developed countries of the North, to socioeconomic advancement, and finally, let us not forget, to research in general. Research is indivisible, and parasitology provides a myriad of fundamental models of large organisms, easily handled, manipulated by the most modern technologies, and capable of producing solid advances in chemotherapy, immunology, vaccine production, genetics, and biochemical pharmacology. In turn these advances would have exciting spin-offs into areas traditionally regarded as the mainstream of pathology, physiology, and therapeutics. For these reasons alone, parasitology, which is entering a phase of expansionist activity, is worthy of our attention. It would be short-sighted of us to allow the looming opportunities to go unheeded or to suffer from neglect.

References

1. World Health Organization (1980) Sixth Report on the World Health Situation 1973–1977. Part I: Global analysis. Part II: Review by country and area.
2. Payne D, Grab B, Fontaine RE, Hempel HG (1976) Impact of control measures on malaria transmission and general mortality. Bull Wld Hlth Org 54: 369–377.
3. Griffiths RB (1978) Veterinary aspects. In: Taylor AER, Muller R (eds.), The relevance of parasitology to human welfare today. Symposia of the British Society

for Parasitology, vol. 16: 41–66. Blackwell Scientific Publications. Oxford, London, Edinburgh, Melbourne.

4. May RM (1977) Food lost to pests. Nature 267: 669–670.

Discussion

WILLIAMS: The point I would have liked Davis to bring up is the relative importance of parasitic diseases compared with the rest of the illnesses in the developing world. How can we arrive at any estimate of how important parasitology is compared to, say, nutrition or bacterial infections? This is the argument one has to fight with nutritionists and microbiologists. One has to be stronger than they are if one believes parasitology is so important.

DAVIS: At most meetings I would not present it that way because I'm eager to arouse more interest in parasitology. In fact, we are doing precious little with parasitic diseases.

To take the example of a director of medical services in an African country who is faced with the difficulty of measuring the economic effects of parasitic diseases, he doesn't have any time to do anything about them because of the importance of ridding the people of their excretions and providing them with potable water and piping: they need pipes for water and for excreta, and pumps for water.

The medical director also has to run medical services that are curative, diagnostic, preventive, and rehabilitative on a very small budget, and provide an immunization service for children and programs for maternal and child health care. And he must also work with the aged and on fatal diseases.

So I am afraid the priority given to chronic parasitic diseases is rather low in most African governments. I don't see the situation changing in the immediate future.

WARREN: Several years ago Julia Walsh and I wrote a paper called "Selective Primary Health Care," in which we compared the major causes of mortality and morbidity in the developing world.

It turns out that the diarrheal diseases and the pneumonias are the greatest causes of infant and child mortality in the developing world. The diarrheal diseases are largely of the bacterial and viral variety. We really don't know the causes of the pneumonias yet, but they in particular seem to be based largely on the debilitation presented by other diseases. Third on the list is malaria. Many of the parasitic diseases are also major causes of morbidity and mortality in the developing world.

I would like to briefly mention nutrition. In his paper, Worboys intimated there was a powerful nutrition lobby; he also indicated that the basic underlying cause of infectious diseases in the developing world is malnutrition, which makes people more susceptible to infections. Some of the latest thinking is that the agricultural lobby is exceedingly effective, and in many areas of the world malnutrition, although a major problem, is not as bad as it once was.

Studies by immunologists suggest that only severe degrees of malnutrition produce significant impairment of the immune system.

Furthermore, a meeting was held recently on the relationship between diarrheal diseases and malnutrition. Work on this problem is going on all over the world—in Gambia, Guatemala, Costa Rica, India, and Bangladesh. The results suggest that probably the greatest cause of malnutrition in the developing world is diarrheal disease, because it is the custom for mothers not to feed their children during these episodes. The resulting caloric deficiency is enormous, and the infants lose an enormous amount of weight; just as they are beginning to catch up, they get another bout of diarrheal disease or another infection.

We are in a contest here, because the amount of money going into research on the control of these diseases is exceedingly small compared to the amount of international money going into agriculture and population control.

The global community has enough resources to support work on all three of these areas in a better way than it is doing. Health is the least well-supported of this triumvirate of crucial issues for the welfare of the developing world.

WILLIAMS: Don't say we are not in a contest, because there is no free money waiting to be picked up. It has to come from some other programs; that is the contest.

WARREN: When one looks at the distribution of money on a global basis, we don't have to take money from agriculture or from population to put into health. We can take money from other things, however, and I don't have to emphasize what they are.

The money is there, but the will to take it is not there on the part of either the developed or the developing world because there is such an enormous focus on diseases of aging, such as cardiac disease and cancer. If one compares the relative importance of these to the relative importance of diseases of the young in the developing world, they don't signify.

NELSON: As an ex-medical officer of health who had to assess priorities in an area of eastern Africa, when I took over the job about 30 years ago parasitic diseases were mainly organisms of low priority. I had to deal with epidemics of smallpox and with lobar pneumonia, diarrhea, and the rest.

Apart from sleeping sickness, parasitic diseases were not a major problem. But there has been a revolution in thinking about them in the past 30 years. For the first time there has been a demonstration of serious morbidity in a small percentage—perhaps by only 5%—of the population infected with schistosomes. But when we take the total figures for the number of people infected, which Davis presented, it is clear that we are dealing with a greater burden of morbidity-caused parasitic diseases than we had imagined in the past.

Much more important probably than even schistosomiasis are diseases such as leprosy or tuberculosis, which receive a great deal of money because an entire community may become infected.

LUCAS: Quite a lot of the epidemiological work done in the past tended to be on the morbidity of individual problems such as leprosy and schistosomiasis. Broad, multipurpose, community-based epidemiological studies are need to provide a better understanding of the problems.

I am impressed by one observation that seems to be consistent: in local endemic areas of malaria in Africa, whenever there is a simple control programme, including malaria control, by whatever mechanism, the infant mortality drops by about 50%. That has come up in various studies.

In one or two programmes people have observed the health of the population in endemic areas for long periods, and when they have refrained from intervening, the infant mortality rates remain high.

These kinds of observations may give us a clue as to the relative importance of these diseases because, as Dr. Nelson said, the figures on direct mortality may not always be apparent. In some endemic areas, people don't suddenly die from schistosomiasis, nor are they diagnosed as having died from schistosomiasis. What we do not know is what this general load of infection is doing to the response to other challenges, for example, to the severity of pneumonias.

Perspectives on Research in Parasitology

5

Parasitology: The Governmental Interest in the United Kingdom

Christopher C. Booth

There is in the United Kingdom (UK) today no specific discipline that can lay claim to parasitology. Although historically the subject was considered to be an important component of tropical medicine, and still is, there have always been histologists, immunologists, and veterinary scientists among others who have pursued interest in parasitology. It is for this reason that at the present time parasitology in the United Kingdom is not the specific preserve of the School of Tropical Medicine but is being studied in a wide range of University Departments or Research Institutes throughout the country.

The Funding of Universities and Research Institutes in the UK

Most work on parasitology in the United Kingdom in so far as it relates to medicine or biomedical science is carried out in universities or research units or institutes. The Department of State responsible for these areas of higher learning is the Department of Education and Science. Like other Departments of State, it must argue its budgetary requirements from Government in competition, and the Department then allocates money, on one hand for university education and on the other for research. So far as the universities are concerned. the allocation is made to an independent body, the University Grants Committee (UGC), which then has the responsibility of sharing out its budget between the independent universities throughout the UK. The total UGD budget for 1980–1981 was

£942,000,000 recurrent and £89,000,000 nonrecurrent. Within the universities there are many departments working in the field of parasitology and there are also within the university sector the two schools specifically orientated to hygiene and tropical medicine (in London and Liverpool), both of which have a substantial interest in parasitology.

Research in the UK is funded from the same Department of State but through a different channel, the Advisory Board for the Research Councils (ABRC). It is worth emphasising that in the UK, medical research is not funded by the Department of Health but by the same body that supports scientific research throughout the academic community. ABRC receives a budget (£383,000,000 in 1980–1981) and is responsible for allocating funds annually to the different Research Councils (Science and Engineering, Medical, Agricultural, Natural Environment, and Social Science) as well as for the country's leading scientific body, the Royal Society, and to the Natural History Museum. In addition it makes a small subvention to the London Zoo. The Medical Research Council (MRC) must therefore argue its case annually for the funds required for its research in competition with these other bodies. If one excludes the Agricultural Research Council (ARC), it is the MRC which has the major interest nationally in parasitology. The total budget for the MRC for 1980–1981 was approximately £93,000,000

The MRC and Its Boards

The MRC is run by a Council to which four Boards with different scientific interests are responsible. The Boards are concerned with Neurobiology and Mental Health, Cell Biology and Disorders, and Physiological Systems and Disorders; the fourth Board is specifically responsible for tropical research, the Tropical Medicine Research Board (TMRB). TMRB, unlike the other three boards, does not advise only on funds allocated by MRC, but approximately 20% of its annual budget is provided by the Overseas Development Administration (ODA), which channels its funds for overseas and tropical research predominantly through the agency of TMRB on whose Board the ODA is represented by a senior official. The total expenditure of TMRB during 1980–1981 was approximately £5.3,000,000, which is between 5% and 6% of the total MRC budget.

Types of Support Given by MRC

The MRC supports research either *directly* within its own institutes or units, many of which have close links with universities, or *indirectly* through programmes and project grants which are predominantly made to university staff. So far as parasitology is concerned, the MRC supports the subject within its own institutes (e.g., the Parasitology Division at the

TABLE 5-1 Numbers of MRC supported projects in different areas of parasitology

Trypanosomiasis	14
Schistosomiasis	14
Malaria	12
Leishmaniasis	6
Filariasis	2
Tapeworms	1
Onchocerciasis	1
Trichuris	1

National Institute for Medical Research at Mill Hill under the direction of Dr. S. R. Smithers) and in units such as the Biochemical Parasitology Unit at Cambridge (Dr. B. A. Newton) or the MRC Laboratories overseas in Gambia (Dr. B. M. Greenwood). In addition, it supports programmes of research carried out by individuals, such as Dr. A. E. Butterworth on schistosomiasis in the Department of Pathology at the University of Cambridge, and there is a wide range of project grants given to individual university scientists working in the field of parasitology.

Subjects Supported on Project Grants

The range of interest in parasitology in UK Research Institutes and universities can be determined by analysing the subjects of various research projects supported by TMRB during 1980–1981; these are shown in Table 5-1. Tryponosomiasis, schistosomiasis, and malaria were the major subjects of research during that year.

Total UK Support for Parasitology

So far as the MRC was concerned, the total support for parasitological research during 1980–1981 was £2,839,168 representing a little more than 3% of the MRC's annual budget. Of this sum £1,993,831 was allocated in direct support of Medical Research Council staff, and £845,287 in indirect support. If one considers the total allocations for research on parasitology in the UK, further support is given by the Overseas Development Administration in a whole range of areas of parasitology through channels other than the ODA annual allocation to TMRB. Support is given, for example, to the Bristol Tryponosomiasis Laboratory (£166,000) and to the Centre for Tropical Veterinary Medicine in Edinburgh (£196,750), as well as to work on insect vectors in the tropics through overseas aid programmes.

Summary

There is substantial support for parasitology in research institutes and universities in the UK, largely funded by the MRC. The Government interest is also illustrated by the support through ODA given to parasitological research both at home and overseas.

Discussion

WARREN: I would like to ask Booth a question for clarification. Your figures for MRC grants don't take into consideration university grants. In the United States, an NIH grant in addition to the researchers' salaries, animal care, and administration also allows an administrative overhead of 50–100%. It is my understanding that MRC grants pay for research only, and that the costs of many other items are matched by university grants.

BOOTH: It depends on whether one is dealing with direct or indirect support: 60% of our funding is direct. So if the MRC is supporting David Weatherall's unit at Oxford, for example, we would expect to pay all overhead. If it is an in-house activity in the university, the staff is paid by us and is responsible to us. On the other hand, when a program grant is given to a university employee, the MRC supports only the research. It is a dual support system, which implies that the university provides a well-funded laboratory, and we provide money for research.

In view of the fact that in the past year university budgets have been cut by 15%, there has been increasing pressure from the universities, and the MRC is giving increasing amounts of support for indirect activities such as animal upkeep. Roughly speaking, direct support costs twice as much as indirect support, so those figures are doubled for indirect support.

WILLIAMS: The support system is changing. It is possible for a foundation such as the Wellcome Trust to pay no overhead on grants to universities. If it did have to pay overhead, its effectiveness in terms of money for research would be greatly reduced.

I fear the financing of universities is moving toward a situation where both the research councils and private bodies will have to pay the additional costs, which will be quite damaging.

BOOTH: We are very much worried about this. I want to see the research base in universities preserved at all costs. Many of us in the MRC would rather see a reduction of direct support and maintain the indirect support because it is an absolute necessity for the universities.

6

Parasitology:
Commonwealth Perspectives

Kenneth L. Stuart

The wide diversity of countries of which it is comprised makes it difficult for an overall assessment to be made of the status of parasitology within the Commonwealth. This diversity makes generalizations impossible and uniform solutions unlikely. But a concomitant of this diversity is that there is hardly a Commonwealth country or region that does not have something to contribute to knowledge about or to the global fight against disorders as widespread as the parasitic disorders.

Commonwealth Organization

Let me say a few general words first about the Commonwealth itself and its arrangements for national and regional health consultation and collaboration among its member countries. Commonwealth membership now comprises 46 countries and approximately one billion people, one-quarter of the world's population. It is made up of countries at all levels of economic development. It is a unique club in which differences rather than similarities characterize the membership. Some of the richest countries of the world and many of the poorest are included. All colours, all races, all religions, all political convictions, all geographic regions are represented. There are certain common features, however, shared by this mix of countries which facilitate collaboration between them. There are, for instance, administrative and civil-service procedures that stem from their past association with Britain, there is use of a common language, and there are national and social aspirations based on shared educational and historical experiences. And all this is enlarged by the facility for frank, friendly, and informal dialogue that is one of the Commonwealth's special advantages.

Probably the most significant of the features they share is a traditional commitment to mutual assistance and support. It is this feature that has enabled the Commonwealth to become such an effective self-help society. It is this feature also that uniquely adapts it for a leadership role in issues that cross national, cultural, and racial boundaries.

The headquarters of the Commonwealth is its Secretariat in London. Through its functional divisions, of which the Medical Division is one, it services meetings of Commonwealth Ministers and expert groups and helps with the implementation of their decisions. Through its Commonwealth Fund for Technical Cooperation, it supports a wide range of programmes throughout the Commonwealth—educational, economic, scientific, health, agricultural, legal, women and development, industrial, export marketing, and others. In these fields it provides two major areas of support. It promotes education and training through the provision of scholarships and bursaries; and it provides technical assistance mainly through the appointment and secondment of experts, for national and regional projects.

The Commonwealth Foundation, which was established by Commonwealth Heads of Governments for the specific purpose of promoting closer professional cooperation within the Commonwealth, has also made major contributions towards these objectives. It has close collaborative links with the Medical Division of the Secretariat. Its trustees have given a high priority to the support of individuals who have demonstrated a willingness and ability to contribute to the grass-roots development of their countries. They have also laid special emphasis on the promotion of activities in the professional field which are not normally covered by larger trusts and aid agencies.

Both the Secretariat and the Foundation attach the highest priority to regional collaboration in all fields. Indeed, for many Commonwealth countries, particularly the smaller and more remote ones, there is no alternative to regional cooperation if satisfactory progress is to be made. In the health field reasonable advance in almost any area will almost certainly depend on the success of the arrangements that can be set up for regional cooperation.

Commonwealth regional cooperation in health has been facilitated through Commonwealth regional health secretariats, the establishment of which was one of the early health initiatives undertaken by the Commonwealth Secretariat. The first of these Secretariats, with its headquarters in Lagos, services the annual meetings of Commonwealth West African health ministers and helps to implement their decisions and to initiate and sustain action on the regional programmes they agree on. Initially including Nigeria, Ghana, Sierra Leone, and Gambia, membership now includes Liberia, and it is now appropriately called the West African Health Community.

We have also supported the development of a regional Secretariat for the Conference of Health Ministers of Commonwealth countries of East, Central, and Southern Africa. Its membership includes Kenya, Uganda,

Zimbabwe, Malawi, Zambia, Tanzania, Lesotho, Botswana, Swaziland, Mauritius, and the Seychelles. Assistance is also provided for the programmes and policies of Commonwealth Caribbean Health Ministers. The formation of a grouping of Pacific Health Ministers and opportunities for their collaboration on a continuing basis are also being explored.

The Health Ministers of the regions hold regular meetings—at least one annually—at which health programmes suitable for regional collaboration are identified and arrangements made for their implementation. The Medical Division of the Commonwealth Secretariat participates in these meetings, making it possible for participants to be kept abreast of relevant problems, projects, and programmes in other countries and in other regions of the Commonwealth.

It is through these arrangements and facilities for regional collaboration that the Commonwealth is likely to make its greatest contribution to the global problem of combating parasitic disorders. None of the parasites or vectors concerned recognise the geographical boundaries that separate countries; and few of the countries affected are in a position to tackle their parasitic problems in isolation.

Difficulties in Fighting
Parasitic Disorders

There would be no point in attempting to give statistical details of prevalence rates of the parasitic disorders in individual countries. I shall merely review some of what we consider to be the more important constraints in the fight against them and some of the ways in which the Commonwealth and the Commonwealth Secretariat might be of assistance. In relation to this assistance, it would be one of the Commonwealth's prime concerns that it should complement and extend rather than duplicate the contributions which are being made by the World Health Organisation, the Rockefeller Foundation, The Wellcome Trust, academic schools of tropical medicine, and a number of other institutions throughout the world.

One of the most critical constraints is probably the low level of political priority that so many Commonwealth governments assign to their national parasitology problems and programmes. Most of the existing programmes are run and managed by external agencies. It is true that these programmes are set up with full concurrence of governments. The gap, however, between concurrence on the one hand and political commitment and responsibility on the other is wide and is only infrequently bridged. In spite of their obvious importance, the parasitic disorders have been conspicuously missing from the list of national or regional health agenda items in most developing countries; and when present they do not lead to discussion in depth. Although this is partly due to the relative weakness of the planning and executive capabilities of national health officials, it is the absence of national commitment and determination that constitutes

82 Kenneth L. Stuart

the single most important barrier to progress in the development of parasitology as a biomedical science of international importance.

A second difficulty, probably stemming from the first, is the overdependence of the developing world on the skills of people from developed countries. That there would be some dependence is to be expected. This is a reality both for the present and for much of the foreseeable future. It in no way justifies, however, the acquiescent assumption that workers from developed countries will always be available and prepared to devote substantial portions of their time to the understanding and solution of the problems of health and disease in the tropics and that the provision of substantial supporting funds from Western organisations and agencies will continue.

A third constraint, also possibly related to the first, is the relative meagreness of the efforts that are being made to connect up scientific advances with the day-to-day concerns and needs of the mass of the people. It is this neglected interface that prevents many of the benefits of scientific achievement from being realised. There can be no question that modern medical technology has been a major factor in influencing health in the developing countries. Scientists alone, however, cannot provide all the answers. Wide community participation, political commitment, and collective national action on an unprecedented scale are required. It is essential that governments, health agencies, and foundations pay more attention to the field team, much of which will be local, and to the critical objective of strengthening its motivation, responsibility, and commitment. Linked to this is the related need for more emphasis by Western Agencies and scientists on the development of capacities for collaboration at appropriate levels with less-skilled workers.

A fourth constraint is the weakness of the health planning and executive capacities of those countries which need them most. Improved capabilities in health planning and management is an essential condition for realising the benefits of biomedical advances in parasitology or in any other health field.

Roles for the Commonwealth Secretariat

There is clearly no simple formula for dealing with these constraints; but there are certain areas in which the assistance of international educational and aid agencies might focus special attention. The areas to which the Commonwealth Secretariat might best contribute are likely to be in (1) strengthening national and regional health and educational institutions—with the aim of producing not only technical experts but field leaders in the relevant areas, (2) promoting regional collaboration in health and in the setting up of an appropriate organisational and administrative framework for achieving it, and (3) stimulating a greater level of concern for the parasitic disorders among national health planners and a higher political priority for grappling with them.

Strengthening Existing Institutions

This is likely to be a major element in any national or regional health programme, and it is to this objective that the Commonwealth Secretariat already attaches the highest priority. Much support is already being given to strengthening existing educational institutions to enable them to become not only regional centres of academic excellence but more effective focal centres for national and regional planning and action in relation to parasitic and other health problems. Our support has been directed to enabling such centres:

1. to help to establish and to coordinate the activities of regional health assistance groups;
2. to assist in the collation and distribution of reference information relating to national and regional health needs;
3. to assist in the planning, implementation, surveillance, and evaluation of educational and other programmes designed to meet perceived local needs;
4. to provide special technical assistance to governments as the need is identified.

Promotion of Regional Collaboration

As a means for helping developing countries to make the most economic use of limited resources, regional collaboration has long been a Commonwealth priority. In spite of the regional health groups that have been established in the Caribbean, the South Pacific, West Africa, and East, Central, and Southern Africa much remains to be done. There is a need for a sharper focus by them on the special health needs of the countries in their regions and for the development of more effective regional programmes for dealing with them. This is particularly true for the parasitic disorders.

The Commonwealth Secretariat is directing major efforts towards helping member countries to achieve a more effective organisational and administrative framework for regional collaboration. Details are being worked out in individual regions, but in each region the Secretariat is giving priority assistance to the promotion of more effective instruments for health collaboration.

Strengthening Planning and Administration

Our close relationships with health ministries and national political leaders enable us more effectively to influence national policies on such matters as the need for strengthening of planning and administrative capacities and for attaching greater priority to disorders which, like the parasitic disorders, compromise the health of such large numbers of their communities. It is for this reason that we are proposing to Commonwealth Health Ministers that Health Planning and Management should be the theme of the next Commonwealth Health Ministers meeting to be held in Ottawa in October 1983.

In view of the goal-setting and policy-formulating role of this meeting, it is anticipated that recommendations based on this theme will provide major Commonwealth emphases for the next triennium. In this context it is also anticipated that we will be able to focus wide Commonwealth popular and political attention on the parasitic disorders and to achieve more determined support for national plans for dealing with them.

Financial Support

As far as financial support is concerned, a number of factors inhibit the mobilization, coordination, and utilization of both internal and external finances in developing countries. These include the complex procedures of aid agencies, lack of information on potential sources of finance, and limited manpower capability for identifying, preparing, and implementing suitable projects. The Commonwealth Secretariat might be able, with its special knowledge of and its relationships with national and international health aid agencies, to help with some of these issues. Although their financial resources are limited, its divisions for general technical assistance and for education and training can be both flexible and early in their response to requests that fall within their financial capabilities and their terms of reference. It is this that has made their support in the health field so effective to date. In view of its global importance, their response to requests for support for the development and application of studies in parasitology is likely to be favourable.

7

Parasitology Research
in the Universities of
Britain and Northern Europe

Bridget M. Ogilvie

In this paper, I have confined myself to the research going on at present in parasitology in Britain and certain countries in northern Europe. It is very much a personal view, and it is by no means a comprehensive or scholarly analysis of the subject. I believe, however, that parasitological research in these countries is in a very active and healthy state and that the major threat to advance lies in the current difficult international and national economic situation which has had and is having severe effects on the ability of the universities to recruit new staff, and thus to renew and replenish the intellectual vigour of their research effort.

Why Study Parasites?

The reasons why scientists study parasites may seem self-evident, but they are worth rehearsing because I believe they strongly influence the nature of the research being undertaken in Britain and northern Europe. The majority of scientists involved in parasitological research especially in Britain are biologists; there are very few medically trained scientists using their clinical skills in parasitological research. Thus, the series of fascinating problems in biology presented by the complex life cycle of parasites are a major influence in determining the nature of the research being undertaken.

The other major motive for studying parasites is a wish to understand and manage the diseases caused by maladjusted parasitic relationships. At present, there are few medical graduates studying parasitic disease, so research into this aspect of parasitology also lies largely in the hands of scientists who are not practising clinicians. In the veterinary schools, however, this aspect of parasitology has always interested considerable numbers of scientists, and current studies emphasize in particular the epidemiology and control of parasitic diseases in domestic livestock.

Where Is Parasitological Research Undertaken?

Historically, and even now, the Schools or Institutes of Tropical Medicine established throughout Europe in the colonial era are major foci of parasitological research. The study of African trypanosomes has always been and remains a special preoccupation of European as opposed to American or Australian parasitology, a consequence of African colonialism. Trypanosomes have preoccupied not only the Tropical Medicine institutes but also many veterinary graduates, and veterinary schools are another major focus of parasitological research in Europe. In Britain especially, there is in addition a very large body of scientists whose research concerns parasitology and who are based in "pure science" departments, especially departments of Zoology or Biology with a very important sprinkling in departments of Biochemistry or Genetics.

The list which follows is not comprehensive but indicates many of the universities and departments in which research into parasites is occurring at present, together with their major research interests.

Germany

BERNHARD-NOCHT INSTITUTE, HAMBURG.
Filariasis—epidemiology, biology, and control of vectors, strain characterisation, chemotherapy
T. gambiense—epidemiology, pathogenicity, strain differentiation
Leishmaniasis—epidemiology, immunology, strain variation
Biochemistry of protozoa

INSTITUTE OF TROPICAL HYGIENE, HEIDELBERG.
Echinococcus
Filariasis
Schistosomiasis

TROPICAL INSTITUTE, UNIVERSITY OF TÜBINGEN.
Biology of onchocerciasis, including vectors
Experimental filariasis
Taxonomy and ecology of snails
Nutrition and malaria

Belgium

INSTITUTE OF TROPICAL MEDICINE, ANTWERP.
Immunology and diagnosis of African and American trypanosomes
Immunology and diagnosis of leishmaniasis
Immunology and diagnosis of malaria
Immunopathology of schistosomiasis
Vectors and pathology of filariasis
Cysticercosis in cattle

INTERNATIONAL INSTITUTE FOR CELLULAR AND
MOLECULAR PATHOLOGY, BRUSSELS.
Chemotherapy of protozoa

FREE UNIVERSITY, BRUSSELS.
Molecular biology of African trypanosomes

Netherlands

ROYAL TROPICAL INSTITUTE, AMSTERDAM.
Schistosomiasis
Ancylostomiasis
Toxoplasmosis

BIOCHEMISTRY DEPARTMENT, UNIVERSITY OF AMSTERDAM.
Molecular biology of African trypanosomes

INSTITUTE OF MEDICAL PARASITOLOGY, NIJMEGEN.
Biology and immunology of malaria

DEPARTMENT OF PARASITOLOGY, UNIVERSITY OF LEYDEN.
Malaria
Toxoplasma
Schistosomiasis
Mosquito behaviour

INSTITUTE FOR TROPICAL AND PROTOZOAL DISEASES, UTRECHT.
Biology and immunology of protozoa

VETERINARY FACULTY, UTRECHT.
Epidemiology and control of veterinary parasites

NATIONAL INSTITUTE FOR PUBLIC HEALTH, BILTHOVEN.
Immunodiagnosis of parasites

United Kingdom

LIVERPOOL SCHOOL OF TROPICAL MEDICINE.
Experimental chemotherapy of malaria, leishmania, filariasis,
   schistosomiasis
Schistosomiasis
Biochemical taxonomy of vectors
Mathematical ecology and epidemiology
Tick and tsetse fly biology
Echinococcus (with Veterinary School)
Pharmacokinetics/genetics of drugs in man (with Department of
   Pharmacology)
Anaemia and malaria

LONDON SCHOOL OF HYGIENE AND TROPICAL MEDICINE.
Immunology of schistosomiasis
Chemotherapy, immunology, and biology of filariae
Immunology, chemotherapy, epidemiology of malaria
Genetics of vectors
Immunology and speciation of African and South American
   trypanosomiasis
Strain characterisation of amoebae
Genetics of resistance to leishmania

DEPARTMENT OF ZOOLOGY, GLASGOW.
Biology and immunology of African trypanosomes
Immunology of malaria
Cestode biology and immunology

DEPARTMENT OF BIOCHEMISTRY, GLASGOW.
Biochemistry of schistosomes

DEPARTMENT OF MICROBIOLOGY AND IMMUNOLOGY, GLASGOW.
Lymphocyte changes in intestinal infections

VETERINARY SCHOOL, GLASGOW.
Immunology and pathophysiology of African trypanosomes in cattle and
   rodents
Epidemiology, pathophysiology, and control of fascioliasis, haemon-
   chosis, and ostertagiasis in sheep and cattle
Analysis of IgE production

DEPARTMENT OF BIOLOGY, IMPERIAL COLLEGE, LONDON.
Mathematical analysis of epidemiology
Echinococcus

Biology of protozoa and their vectors
Hyperparasitism

UNIVERSITY OF CAMBRIDGE.
Molteno Institute—biochemistry and molecular biology of protozoa
Department of Pathology—schistosome immunology
Veterinary School—immunology of veterinary helminths

UNIVERSITY OF EDINBURGH.
Departments of Genetics and Zoology—genetics and strain variation in
protozoa
Department of Molecular Biology—molecular biology of *Plasmodium
falciparum*

Another special aspect of research in Britain is that much of it is carried
out outside the universitires in the research institutes funded by the Medical Research Council and the Agricultural Research Council, and the
activities of these organisations are therefore not specifically mentioned in
this review.

The Institutes of Tropical Medicine in Europe have links with other
countries, as follows:

London School of Hygiene and Tropical Medicine—Sudan, Brazil
Liverpool—Ghana, Bangkok
Antwerp—Bolivia, Zaire, Camerooon
Leyden—Surinam
Hamburg—Liberia ⎫
Tubingen—Nigeria ⎬ laboratories managed from Europe
Amsterdam—Kenya ⎭

Major Research Themes at Present

To obtain an up-to-date picture of the nature of research at present going
on, papers presented at scientific meetings were analysed. There are societies devoted to parasitology in almost all European countries, and parasitological research is also reported at, for example, meetings of the Societies of Tropical Medicine at veterinary meetings and at many specialised
societies. The largest society devoted to parasitology in Europe is the British Society for Parasitology. Although a young society by British standards,
as it is less than 20 years since its inaugural meeting, it has about a thousand members. It also has strong links with certain European countries
and in 1981 its major meeting, which is a three-day residential affair held
annually in April, was held in Holland jointly with the parasitology societies of Belgium and the Netherlands. These meetings of the British Soci-

TABLE 7-1. British society for parasitology: research topics

	% of total papers	
Subject	1981	1982
Immunology	24	29
Biochemistry and physiology	10	14
Biology	7	21
Epidemiology	7	9
Ecology	10	9
Molecular genetics	0	8
Chemotherapy and control	13	3
Taxonomy	10	3
Pathology/pathophysiology	7	3

ety for Parasitology have two especially impressive characteristics, a very high attendance—in 1982 there were over 400 attending—and the youth of the participants especially in comparison with, for example, the Royal Society of Tropical Medicine or the British Society for Immunology. The papers given at the 1981 and 1982 Spring meetings were analysed by topic (Table 7-1).

Taking immunology as the example, a more detailed analysis of the papers showed that parasitology research is becoming more and more analytical in its approach (Table 7-2).

TABLE 7-2. British society for parasitology: immunology research

	% of total papers	
	1981	1982
Intact host and/or parasite[1]	40	44
Analysis host reponse by Ab class, or cell type	10	28
Analysis of antigens[2]	35	23
Diagnosis	15	5

[1]Includes comparisons of host strains, stage specificity, kinetics of response, immune avoidance, concurrent infections, immunosuppression, immunisation with whole parasites or undefined products.
[2]Definition of antigens, monoclonal studies.

Future Research in Parasitology

As indicated earlier, the strength of European parasitology lies in its roots in studies of the biology of parasites and the natural history of the diseases they cause. A good illustration of this is the current preoccupation with variations occurring within previously established parasite species linked with the nature of the disease each subspecies induces. For example, recent work on *Echinococcus granulosus* has made us realise that there are probably several subspecies of this parasite which vary in their host specificity in both the larval and adult stages of the life cycle. Other studies suggest that we do not know whether *Onchocerca volvulus* exists as a single species or whether there are several subspecies, as suggested by careful studies of disease patterns in different regions. Again, the present flowering of research into *Leishmania* is very dependent on still fruitful biological and epidemiological studies of the parasite, its vectors, and the lesions caused in man. The strain variation recently discovered using isoenzyme analysis of *Entamoeba histolytica* is being related to the pattern of disease found in man. These variations in parasites within species are immediately open to analysis at the genomic level using modern techniques, but the diseases that may result are largely neglected. There is at present a major preoccupation concerning the need to apply the techniques of the "New Biology" to parasites, and there can be little doubt that these powerful genetic and immunological tools will be revolutionary in their impact in this as in other aspects of biology. However, to me an equally important and extremely neglected aspect of parasitology is the need for more research into the clinical management and pathophysiology of diseases caused by parasite infections. Here advances can still be made using well-established methods as well as recent innovations, and the major deficiency seems to be a lack of clinically qualified research workers. Veterinary research leads the way here as illustrated by studies that show that genetic variations in the host influence the severity of the disease induced by parasites. Careful field studies and pathophysiological analysis of the pattern of disease has shown that some breeds of cattle and sheep are better able to resist the effects of parasites than others. Comparable studies of disease induced by parasites in man are few, especially when compared with the explosion of interest in parasites involving laboratory-based studies. Our understanding of the disease induced by malaria and its clinical management has advanced little at a time when knowledge of the antigens of malaria is advancing rapidly.

For these reasons, I feel it is necessary to preserve a balanced view so that in our excitement at the promised advances indicated by the application of new techniques we do not forget that much can still be learned by the imaginative and skilled application of tried and tested methods. To illustrate, studies on the epidemiology of *Leishmania* which relate the disease pattern in man to the reservoir hosts, vectors, and parasite subspecies involved are just as important as studies on the genetics of the host response using mice or the development of monoclonal antibodies and DNA probes to investigate speciation and antigenicity. My own recent research was dependent on extensive collaboration between individuals with different skills. Without colleagues able and willing to maintain the life cycle and undertake experimental studies of filarial nematodes, colleagues skillful in handling cells and parasites *in vitro,* medical colleagues in areas of endemic filarial infection, and the biochemical skills of other individuals, little progress would have resulted. In all aspects of biology today, real advances usually require the simultaneous application of many skills, and progress depends increasingly on successful collaboration. In any aspect of science, however, if understanding is to be increased, nothing can replace hard work by imaginative and observant scientists, whatever techniques they may employ.

Discussion

WARREN: I agree completely with what Ogilvie had to say about support for the full spectrum of parasitology.

The thing that worries me is the financial situation in the United States. At the moment there seems to be a relatively constant pool of funding for parasitology from the NIH; we have a large cadre of classical parasitologists, however you interpret that term, who are doing good work; and there is an influx of new people coming into the field, people with a great potential for important work. The same thing is going on in Australia and in France.

Given the financial situation in most parts of the world, however, it seems to me that there is going to be an enormous financial squeeze, and one side or the other is going to lose drastically if not enough money is put into support for the new biology; if it is not, that very crucial movement will be nipped in the bud. If at the same time that takes funds away from classical parasitology, then a significant basic underpinning of parasitology and of the new biology will disintegrate.

We cannot sit back and say that financing is adequate in any of our countries and that it is rising slowly year by year. If we are going to take full advantage of the opportunities afforded now, we have to find ways to increase funding for this field. I believe Capron mentioned that the total funding for parasitology in France is 1% of the total research budget. That is about what it is in the United States.

BOOTH: In Britain it is about 3% of the MRC budget.

I think we are on very difficult ground if we take any specific field of science and say that because something new is happening in it more money has to be invested in it. The reality of the situation in Western Europe unquestionably is that research budgets are going to stay the same or even go down. That is a fact of life. It is also a fact of life in the case of the NIH. Overall I suspect that it's going to be true of all research budgets.

If that is the case, there is no question that if we want to do something new with research funds, we have to stop doing something old. That means we have to be able to translate the old into other items in another budget—perhaps standard university salaries or routine government work—that relates to whatever we are doing. The other side of the equation is the translation of existing research into routine structures.

WARREN: I think you are right, but don't think we should become resigned to that fact. We must all fight for higher budgets because we are competing for money for research in other areas, such as military, agricultural, and population control.

When I see the biomedical research budget in the United States remaining constant or being cut, while the military research budget soars, and everyone says, "This is fine," knowing what a great opportunity there is, I know we have to do everything we can.

We may have to do exactly what Booth says, but we still have to apply as much pressure as possible to get more funds for the area.

WILLIAMS: If we really want to sell the situation on a priority basis, we have to make more opportunities for practical advances. I am bound to say the pool of money is fairly constant in this situation. There may be pressures we can bring to bear, but it will turn totally on new prospects for practical advances. That is what people give money for.

SELL: The opportunities that are raised by the new tools provided to us in this field are tools that can be taken up by anyone in any field. They are being used by cancer specialists and investigators working on heart disease and on metabolic diseases, but most particularly in other areas of infectious diseases, such as vaccine production.

The enthusiasm for doing work in this area should not be neglected, but it has to be realized that the new tools are going to create the same thoughts in the minds of those who pursue other areas of research. That should not be overlooked.

These investigators are maintaining the same pressures to try to increase funding for these fields. The difficulty with that is that, overall, everyone is saying, "We have a new opportunity." As has been stated, it is unlikely that there will be a major new source of funding, although as we have heard, there is a continued increase, though modest, in funding in some countries such as France.

In the United States, as in the past, we have had rough periods, and I anticipate in years to come we will have other growth periods.

I am quite certain, however, that the American public will, as in the past, give favored support to biomedical research. Our state of growth will be faster than other elements of government—that has been true, and I am sure it will be in the future. But that rate of growth as it translates into work in parasitology has to deal with the opportunities compared to other opportunities in other areas of biomedical research, which are also being investigated and supported by groups prejudiced in favor of their specialties.

GITLER: The time is ripe now for modern biology to look at complex organisms that we approached before, such as bacteria, at a time when tools were not available. People need to get involved with and at the same time try to solve problems bearing on public health.

The modern biologist will search out whatever systems allow him to ask the question he wants to ask. The best case in point has to do with the system of transformation between an amoeba and a flagellate. For a long time it seemed to make sense only from the morphological point of view. Today it is a fascinating development, and it is probably the most accessible system available for any developmental biologist to study the expression of the genes being triggered simultaneously.

I have no doubt that if made available to modern biologists the system will generate solutions to malaria and malarial encephalitis. Why should we invent problems in sophisticated science when today we have actual parasitological problems relevant in both areas? In other words, they are relevant both in practical and in theoretical ways. If one emphasizes the practical aspect more strongly, sophisticated people will not be attracted unless one can combine the practical with the asking of important questions.

WILLIAMS: Why shouldn't sophisticated people work on practical things?

8

Research in Parasitology in French and Overseas Pasteur Institutes

André Capron

Historical Background

Parasitology Research in France leans on a long tradition which, for more than one hundred years, has been continuously maintained in Research Institutes or Universities. The unique position of Pasteur Institutes in France and Overseas in this area of Medical and Veterinary Research certainly finds its origins in the pioneering work of Pasteurian parasitologists.

Although it is not the primary aim of this contribution to give an historical overview of parasitological research in France, it seems appropriate, as an introduction to present structures and orientations, to call to mind some of the illustrious names which have marked the first steps of French parasitology.

A few years after his discovery of malaria parasites, Alphonse Laveran left the army as a colonel and, on the invitation of Roux, entered the Pasteur Institute in Paris, where he worked until his death in 1922. He first found a welcome in Metchinkoff's laboratory, then in 1907, thanks to the Nobel Prize funds that he had just been awarded, he created a specialized department which included three main laboratories (protozoology, microbiology, and medical entomology), respectively led by Mesnil, Marchoux, and Roubaud, whose work on trypanosome and leishmania biology and on the first therapeutic trials on sleeping sickness laid the foundations of modern protozoology, which were later built on by Caullery, Chatton, and Lwoff.

At the same time, at the Pasteur Institute of Algiers, the two brothers Edmond and Etienne Sergent initiated their work on malaria, which led to the definition of premunition immunity, a brilliant anticipation of the concept of concomitant immunity. Also at the same time, Charles Nicolle, in the Institute of Tunis, accumulated major discoveries: infantile kala azar, the mechanisms of transmission of canine leishmaniasis, the mechanisms of transmission by lice of exanthematic typhus, and the discovery in a wild rodent of a parasite named *Toxoplasma,* the pathogenicity of which is now considered a public health problem in many countries.

The creation before World War II of Pasteur Institutes all over the world allowed the building up of an active network of international research centers in parasitology which have greatly contributed to field studies as well as to basic research on the taxonomy and biology of parasites or vectors. It is noteworthy that during the same period, Emile Brumpt, at Paris University, developed a very important group and became one of the international leaders in parasitology.

The consequences of the Second World War and the progressive access to independence of many countries in which Pasteur Institutes were settled, the restructuring of French research in several major organisations— INSERM (Institut National de la Santé et de la Recherche Médicale), CNRS (Centre National de la Recherche Scientifique), ORSTOM (Office de la Recherche Scientifique et Technique Outre-Mer), INRA (Institut National de la Recherche Agronomique)—have led to the present stiuation which we will now analyse.

Present Research in Parasitology

One of the main characteristics of the French research system is certainly the diversity of the public or private national agencies, claiming their responsibility for managing specific programs. Parasitology as such is not really the subject of a conflict of interests, all the more that for over 30 years the main national research organisations (INSERM and CNRS) have obviously shown little interest in supporting parasitology programs. The situation has now slightly changed, since medical research for developing countries a few years ago became one of the political priorities of our country. As an important component of research directed towards the Third World, parasitology has progressively moved from a marginal to a rather central situation in medical sciences. Even now, however, financial support for parasitology remains rather theoretical. The support of INSERM, for instance, to parasitology hardly exceeds 1% of the total medical research budget. Of more than two hundred INSERM Research Units, only two have a real research specificity in Parasitology. The same comment could be applied to the CNRS.

It is in this particular context that the Pasteur Institutes decided six years ago to reinforce their research potential in this field. One of the first concrete results was the constitution in 1975 of a Centre for Immunology of Parasitic Diseases, at the Pasteur Institute, in Lille. This group, which now comprises 35 scientists and 55 research technicians, administrative personnel, and animal technicians, is located in a modern building of 1700 square meters of laboratories and 1000 square meters of animal houses.

Three parasite models are at present investigated: schistosomiasis *(S. mansoni)*, which represents the major field of activity, filariasis (including experimental infection by *Dipetalonema viteae, Brugia malayi* and human onchocerciasis), and American trypanosomiasis *(T. cruzi)*.

For each of these models, a combined approach is directed towards basic studies on immune mechanisms and parasite biology, and towards applied immunology. The main scientific programs deal with the host–parasite interface surface immunogens, effector mechanisms of defence, and the regulation of the immune response. In addition, goal-oriented studies include the production of monoclonal antibodies against surface and circulating antigens, DNA recombinant technology, the development of new technologies for diagnosis purposes, and the pharmacological investigation of parasite-derived immunomodulators. A close interaction is developed between each of these research components. The main achievements during the past five years have been:

1. the discovery of surface receptors regulating the host–parasite interface; the demonstration of ecdysteroid hormone production by schistosomes in relation with outer membrane and sexual differentiation;
2. the identification using conventional and monoclonal antibodies of surface target antigens involved in protective immunity. The isolation of parasite mRNA and its *in vitro* translation as a first step towards gene cloning has been recently obtained;
3. the identification of ADCC mechanisms as a major component of parasite killing leading to the demonstration of the involvement of anaphylactic antibodies in protective immunity. These studies have led to the description of new ADCC systems and the characterization of Fc receptors for IgE on mononuclear phagocytes and eosinophils;
4. the demonstration of parasite-derived immunoregulatory factors regulating lymphocyte macrophage, eosinophil, and mast cell function;
5. the production of a variety of monoclonal antibodies against *S. mansoni, O. volvulus, T. cruzi,* etc., has allowed the demonstration of the protective role of defined isotypes confirming *in vitro* observations and the development of immunoassays for the detection of circulating antigens at a preclinical stage infection.

This research activity is supported together with funds of Pasteur Institute by INSERM, CNRS, WHO, UNDP Special Program for Tropical Disease Research, and Edna McConnell Clark Foundation.

A few years ago, two new units were opened at the Paris Pasteur Institute. The Immunoparasitology Unit, is mainly oriented towards molecular biology and the immunology of protozoa. The main orientation of this group has concerned, in the last few years, the characterization of surface antigen of *Trypanosoma cruzi*. The demonstration of the so-called fabulation process by which parasite proteolytic enzymes cleave bound antibodies, which can then no longer activate complement and protect the parasite against intact antibodies, has been recently made. This process seems to be related to the variable virulence expressed by different strains.

An important program is now pursued on the molecular biology of trypanosomes. Using *Trypanosoma equiperdum* and *T. brucei* as models, the group has prepared cloned cDNA and studied the expression of genes coding for variant antigens.

The Experimental Parasitology Unit is developing an important program on the immunology of malaria. One of the most important advances has come from the elegant studies made in collaboration with the Pasteur Institute in Cayenne (French Guyana). Using a novel experimental model, the squirrel monkey *(Saimiri sciureus)*, this group has been able to establish an active experimental infection with *P. falciparum* and to study the protective role *in vivo* of transferred antibodies of defined specificity. Immunoprecipitation studies have led to the characterization of several proteins, which are at present under investigation.

Another group, oriented towards biological control of vectors, has made significant contributions in the study of pathogenicity for mosquitoes and other insects of *Bacillus thuringiensus* H14. At the Pasteur Institute in Lyon, a group of pathologists is leading a program of ultrastructural studies on pathogenesis of liver fibrosis in human and experimental *S. mansoni* infection.

In order to be complete, in this brief overview of parasitology research in the French hexagon, one should mention, outside the Pasteur Institute, the research done in various Institutes: biochemistry of malaria parasites in Lille, comparative helminthology and biology of *Plasmodium* in Paris, ecology of *Leishmania* in Montpellier, and biology of schistosomes and their intermediate hosts in Perpignan. One should also mention that many university laboratories in Paris, Grenoble, Strasbourg, and Bordeaux have a continuing activity in the epidemiology and serology of parasitic infections.

The important epidemiological activities of ORSTOM have recently led to significant contributions to the control of vector-borne parasitic diseases (African trypanosomiasis, onchocerciasis, etc.).

Parasitology Research Overseas

It is evident that the major political changes that followed World War II deeply affected the existence of the Institutes in countries which became independent. The important network that existed in former French colonies is now reduced to a few Institutes which are, besides the local administration, coordinated by a central direction in Paris (Direction des I.P.O.M.).

It is interesting to note that, apart from the Institutes in French territories which have maintained a significant activity in parasitology research like Dakar, Cayenne, and Pointe-à-Pitre, very close links have been kept with former colonies like Algeria, Tunisia, and Madagascar and with Institutes in foreign countries like Greece. It is also worth mentioning that two non-Pasteurian institutes are now associated with this network, Institut Louis Mallardé in Papeete (Tahiti) and Institut IBBA in La Paz (Bolivia).

Research in parasitology in these various institutes has been recently restructured and focused on three major areas: malaria, leishmaniasis, and filariasis. In malaria, two main programs have been developed: one in the Cayenne Pasteur Institute where after several years of effort, following the initial discovery that the squirrel monkey *(Saimiri sciureus)* could be experimentally infected with *P. falciparum* and *P. vivax*, a large monkey colony is now established. Numerous studies on the biological parameters of *Saimiri*, their immune response to *Plasmodium* infection, and serum passive transfer experiments have led to the definition of a new and imprtant experimental model for malaria research. Another program, mainly directed towards the study of the immune response in man and the development of seroepidemiological methodology, is carried out in Dakar. These two programs are in close interaction with the Experimental Parasitology Unit in Paris.

Research on leishmaniasis carried on in Dakar, Cayenne, and more recently La Paz has led to important epidemiological observations on vectors and reservoirs as well as biochemical and enzymatic studies on strains and extensive investigation on the cellular and humoral response in human infection. Studies on *Wuchereria bancrofti,* mainly focused on the investigation of the immune response in human infection and on antigen characterization, are performed in Papeete. Epidemiological and immunological studies are carried out on schistosomiasis in Madagascar *(S. mansoni)* and in Algeria *(S. haematobium).*

It appears therefore that parasitology research in overseas Institutes is far from being negligible and that the existing structures certainly represent, if correctly connected to the Institutes in France, a real potential for operational research.

Future of Parasitology Research

After a brilliant period of activity marked by numerous important discoveries, after World War II the French School of Parasitology had some difficulties in defining an adapted research strategy and, in particular, in stimulating the introduction, on the traditional ground of parasite morphology and biology, of the new tools of biochemistry, immunology, or molecular biology. This had, in some ways, left parasitology, within the French research systems, in a rather marginal situation until the year 1960. The scientific orientation, based on modern concepts and tools of research, has progressively changed the attitudes towards what was commonly considered an obsolete discipline. The recent priority given by governmental authorities to research for developing countries together with the recognition of the significant contributions made by several French groups greatly contributes to a less pessimistic view for the future. Scientific authorities are now convinced that a particular effort should be made for the funding of research which is central to many public health problems in developing countries. It is not unlikely, if the economic situation allows it, that an increase of support for the existing structures, together with a better coordination of research, can be expected in the future.

Research, however, is not only a problem of funding. It is also a problem of structures and of individuals. This requires a lucid policy of research organisation and training at the national level. France certainly has the opportunity, because of the close links that are maintained with former colonies or foreign developing countries, not only to maintain but also to promote, in interaction with national institutes, an active research in parasitology. The limits of such a policy are undoubtedly the limits of France itself to maintain research programs adapted to the importance of its ambition to increase cooperation with developing countries.

The perspective of European programs, sponsored by the European Community and aiming at a better coordination of the research efforts directed towards the Third World, can certainly contribute to increasing not only the efficiency of the programs but also, by a cross fertilisation process, the overall quality of research in parasitology.

Discussion

CAPRON: I have several comments to make about what has been said here concerning the function of international organizations in general. We are all impressed by the results and achievements made by the various organizations, but I feel a bit frustrated that no attempt was made in any of the presentations to contribute to an increased potential to develop parasitology research in general. By that I mean an international organization has in fact a dual function.

I would place research as the primary function. This is not only a problem of funding; one of the main objectives of an international organization is certainly to increase coordination and interaction between the various research institutes and laboratories, in general, without any financial consideration. In this respect the TDR Program [Research and Training in Tropical Diseases] is of tremendous importance.

Having been involved in that program since the beginning, one of the major benefits I have personally observed—and I am speaking as a French scientist—has been the strengthening of the links between French research in parasitology and the general trends of international research in parasitology and in tropical medicine.

The second function of an international organization is of course to fund defined programs.

If one takes France as an example, it is obvious that even though the plan is to increase the research budget significantly in the years to come, that increase will cover only current expenditures. A 15% increase in the budget merely keeps up with the annual inflation rate, which allows us to maintain existing activities; it doesn't allow us to start new programs or to bring new people into the laboratory.

One of the main functions of laboratories of an international organization operating in diveloped countries—and again I refer to the effect of the TDR Program on the laboratories operating in Europe, for example— is to provide opportunity over a given period of time to fund special and very specific programs and to hire the appropriate scientists. This is a tremendous impetus to doing something really concrete in maintaining the research potential of parasitology and increasing it in the future. We have to take these two points into serious consideration.

WARREN: Obviously the TDR has accomplished much. How would you put that proposal into action?

CAPRON: I have been working with TDR for many years and know exactly what kinds of interaction and support that program offers, even though it has limited funds. It has been extremely helpful, not only in initiating research, but in coordinating it and in giving responsibility to many scientists.

European countries, and France in particular, always feel a little isolated because the numbers of laboratories and scientists working in this particular field are rather limited.

WARREN: You are absolutely right. It is obvious that TDR does have a major function and that it is a relatively large program. Our program is one-tenth the size, in terms of the financial outlay of TDR. We have 14 units, but we can't expand beyond that, which is a matter of concern to me.

Our meetings involve about 80 people in those 14 units, and our great worry is that, whether or not we expand to include more people, we can't finance more units.

Possibly we could accommodate individuals at our annual meeting, which is a lot of fun and involves a lot of collaborative interaction.

I would be interested in the reactions of some of the other people in this particular network to that proposal. I think, given the closeness, the size of it, that is about the maximum we should have to ensure reasonable collaboration, because the meeting lasts three and a half to four full days, and we have to allow time for everyone who participates to speak. I, for one, would be delighted to have representatives from this group attend. I am trying to think about linkages and networks.

The question is one of practicality, however. How do some of the others in the Great Neglected Diseases Network feel about this? Is it possible to expand?

CAPRON: That is not the point. My strong feeling is that we need better coordination and interaction between the research centers in Europe and in the United States. We certainly need, also, better coordination between international organizations that support research. What we need is to make the organization of parasitology research throughout the world as solid as possible.

WARREN: That question is how to do that in a practical way.

MITCHELL: This is an appropriate time to bring up that issue because, as you said, it is clear that there are financial constraints on how much the Great Neglected Diseases Program can expand.

You have a specific amount of money in the kitty and there are a certain number of groups to support. The way of expanding the program's influence is of course to link up very closely with other national organizations such as those Capron referred to, or organizations such as the TDR and the Wellcome Trust. Their tentacles can go much further than can the GND Program, with its limited funds.

WARREN: By what practical mechanism can this be done?

MITCHELL: It already exists. People such as myself are funded by multiple sources, including the GND. We don't receive Wellcome money, but we do enjoy TDR money. Linking up with people who are in both situations certainly tends to extend things.

Capron's point is that the TDR has been in a way the salvation of many laboratories that are not yet at the cutting edge, but that could be. The French are not in that position, but the TDR program has enabled other countries in the world to strengthen their research activities. There is no question about that.

The issue is how can the GND become more closely associated with the TDR? That is the issue Capron was raising.

CAPRON: Yes. I think we all despair of the very poor methods used by different organizations. It is necessary that we reach a stage where there is a better coordination between them.

ARNON: I don't understand the connection between the funding issue and collaboration. I would gladly collaborate with anyone working on the same topic as I am, whether it is funded by the TDR, by Wellcome, or by the GND.

WARREN: How does one bring them together?

LUCAS: To inject a personal note, I am one of those who receives funds from many sources.

WILLIAMS: There has been no great problem between Lucas and me about work in a place in southern Nigeria that wasn't mentioned in my talk. It has been very successful.

LUCAS: Thank you. That is important because it illustrates some of the things private foundations are doing—training grants from the Rockefeller Foundation and research grants from Wellcome and from Rockefeller. These have made it possible for us to do the research we did in Ibadan on schistosomiasis, for example.

The point that should be raised, however, is the one that perhaps links up with what Capron has said: that is, each organization, and each sector within each organization, has its opportunities and its limitations.

We can do things in the TDR area, for example, that others cannot do so easily, whatever the reason. We have access to some countries, whereas others don't. On the other hand, there are things other organizations can do that we cannot. Rockefeller can give funding over an eight-year period and say, "Do what you like, but do it well." We cannot do that because of the way we are set up.

My friend and colleague, Davis, has more flexibility in using funds than I have in the WHO. So we are looking for ways we can maximize the opportunities available in these various organizations, with each one of us using what we have to the best of our ability in collaboration with others.

From my point of view, the collaboration has been successful. I rarely do things that Warren doesn't know about. We are, for example, regularly invited as TDRs to attend the GND meetings. Dr. Davis has been closely involved with the Edna McConnell Clark Foundation.

The Wellcome Foundation and the Wellcome Trust gave money to TDR at the beginning and, in addition to the official government representative, Dr. Ogilvie and Dr. Williams have attended the TDR meetings as representatives of the Wellcome Foundation.

These activities are valuable and we should look for more opportunities for them.

I would like to see even more cooperation with some of the institutions in developing countries. I would put a very high priority on the issue of strengthening institutions in endemic areas; institutions which are not yet ready to take over the activities that have been going on elsewhere. Unless we start to do that now, they never will be ready.

I would be happy to join my colleagues in cooperating in this particular venture of strengthening institutions. I don't think they should worry about funding the same institutions we are supporting. In fact, we would very much welcome that. Our hope is to put in a little and expect others to contribute a lot more.

BOWERS: The French situation is unique in that France determined many years ago to treat its former colonies almost as part of the empire—and I say this candidly—and it has been extremely effective.

WARREN: Capron has raised a very important point, and I for one am taking it very seriously. We are going to have to see how it evolves during this discussion.

WILLIAMS: I find it fascinating at an international meeting like this to contrast the methods of funding in different countries. France has never developed the concept of private foundations. It has something to do with the system; the French believe their government should be responsible for funding programs. To some extent France suffers from this lack of philanthropic organizations because there isn't that flexibility the United States has in excess. This results in a special type of situation where, when a research group wants additional money, it has to look overseas for support.

In Britain we are about halfway between. Private funding is significant, and as chairman of Medical Research Charities I know they receive some £55 million a year; the Medical Research Council's budget is £105 million a year.

In the United States, while the government contribution to research is obviously enormous, and in much greater proportion than in Britain, the role of foundations has become a very significant developing force in both countries.

I am also chairman of the Hague Club, the meeting ground of the chief executives of the major European foundations, which are involved in various fields. Each of our countries has a different culture, and because of these varieties when we get together to try to take a global look at one subject, such as parasitology, we don't quite understand one another. The United Nations operates in yet another way, of course.

This is a matter of responsibility. As director of a foundation for many years, I consider it irresponsible to start a program unless one intends to continue to support it. I do not think it is proper to invest a sum of money unless one believes either that the program will come to completion at the end of a certain period or that some organization with more money will take it up because it has proved to be so good. One therefore must see a way out at the end or be prepared to go on. If one supports a project, only to let it flounder at the end of, say, three years, then what is the point of doing it at all?

I believe foundation money is by definition short-term money; it is small money, and therefore it is incentive money intended as start-up support. But one must have thought through one's creation.

In the case of sickle cell anemia, the Wellcome Trust supported a young man in Jamaica named Graham Sargent, who worked for three years on a fellowship investigating that disease. He did so well that at the end of that time he approached the Medical Research Council, which gave him a unit. Each project is like that: It has to prove itself.

What I was hoping to bring out today is that, for example, the long-term objective of the Wellcome Trust is the promotion of medicine in the tropics, not parasitology, per se; the objective of the Rockefeller Foundation is parasitology, and modern biology in relation to it; the objective of the TDR is to find tools for the betterment of health in the tropics. Each of these objectives has an interrelationship, but they are not the same. We therefore have to realize that we have our different objectives.

Each of the three programs that we have heard about here—the TDR, the Wellcome Trust, and the Rockefeller Foundation—can be considered as a series of interacting rings, like the five-ring Olympic games symbol, in which there are intersections. The Edna McConnell Clark Foundation is perhaps the fourth ring. There are other financial rings, but there are also intersections. They are different, and that is good.

On the otther hand, I believe Capron has a point; is there some way we can put all this together and maximize communication and collaboration?

SOPRUNOV: We should not forget that, after all, the objective of the programs of the TDR, the Wellcome Trust, and the Rockefeller Foundation is not to promote science, parasitology, or the pharmaceutical industry, but to help people in the developing countries. To me the more interesting of the programs is Warren's because it is closer to the medical problems of the developing countries.

There are three points I would like to make:

1. Is there a hope of improving the condition of populations of the developing countries in the next five years? If we agree the answer is "no," we must realize we have lost our war against the six great diseases.
2. My second point is that although a network of laboratories has been set up in the developing countries by Wellcome, the TDR, and Rockefeller, there is a gap between the laboratories' programs and the practical needs of the ministers of health in those countries. If we ignore that, we make a great mistake.
3. The third point, Dr. Warren, is that you emphasized the great neglected diseases, but there is a great neglected population living in the rural areas of developing countries that we don't take into account. Do you really believe we from other nations can overcome the problems of these people? I believe they can be resolved only by the people themselves. We have to help of course, but we cannot make those problems disappear.

NELSON: I will be very brief because our friend from the Soviet Union has said exactly what I wanted to say. Let me add one thing in relation to what Capron said about the fact that scientists can't get together. They do get together: there are almost too many meetings of scientists in the same area. Scientists communicate, but what we are not doing—and this is not the fault of the TDR—is accelerating progress on the control of the diseases we are working on.

With our knowledge of modern biology we in fact tend to use parasites in the laboratory as beautiful models in the interests of general medical science, and to forget that we used these models originally to try to benefit people in the developing countries.

STUART: Was the question directed specifically to parasitology, Dr. Warren? Williams referred to your interest in parasitology, and that is why we are here.

My first point, therefore, is whether there isn't an intrinsic weakness in separating parasitology in such a clearly defined fashion from the spectrum of other health needs of developing countries. Shouldn't we take a bit of emphasis off of the disparate character of parasitology and seek as our next move to promote it in relation to the health needs of the developing world?

WARREN: I couldn't agree with you more. It is my own belief, which I have expressed many times, that parasitism is a mechanism of all infectious agents. The reason we asked Worboys to participate was to find out why, historically, parasitology became split off from the other infectious disease agents, because not only has it become a separate discipline, it has had enormous repercussions.

Bacteria, viruses, fungi, protozoa, and helminths have all been studied in certain laboratories. My feeling is that if parasitology had remained part of a discipline called parasitism, and if it had been pervasive in medical schools and in research institutes, perhaps it may not have become as isolated as it has and fallen behind the modern advance of biology.

I agree completely that among the diseases of the tropics, the diseases in the so-called discipline of parasitology play a very important part. But the bacterial and viral diseases are at least as important, and in some cases more important. That is why the GND Program includes all the great neglected diseases, including diarrheal diseases and pneumonias, as well as measles. The Wellcome program goes beyond that to deal with other problems in the tropics such as hypertension and asthma, which is a major one.

I think you are right, Sir Kenneth; we are trying to cover a very specific field, which somehow or other has become isolated from the modern march of medicine. We have to do something to improve the status of this very important scientific area and integrate it with the others.

STUART: My second point relates to what Soprunov said. On my visits to developing countries I have noted that their ministries of health have a common channel through which any major improvement in health is going to come about, no matter who does what in which country; it is through these ministries that progress will be made.

Both Warren and Williams have presented maps of their respective, very widespread networks of highly sophisticated and well-financed, important work. My question to you both is: To what extent do the activities represented in those maps tie in with the perceived concerns of the health ministries of the countries in which they are located?

WARREN: Let me say briefly that we are concerned through the program (epidemiology) I mentioned. Second, at the present time we are working on a multiauthor book directed to ministries of health.

Most textbooks on tropical medicine focus on the physician treating an individual patient. We believe it is exceedingly important that a totally different strategy be developed for a minister of health dealing with a population of patients.

Our book is therefore on that specific subject and covers all the so-called major tropical diseases, including viral and bacterial diseases. It has been written by experts in each field, and its purpose is to try to advise ministers of health on the best and most cost-effective way of dealing with each problem.

The issue you raised, Sir Kenneth, is one we are deeply concerned with and believe to be very important.

WILLIAMS: I think it has to be accepted that we were not set up to do what ministries want, and I don't think we should necessarily do what they want. If every organization turns toward the practical issues of implementing new knowledge—which for the most part means doing what the ministries want—we will have a stultifying period during which it will not be possible to move ahead.

I therefore think it appropriate that the small sums of money we give are not necessarily given to develop what the developing countries envisage as their immediate needs. One may perhaps have to go a little further than that and say, "We are thinking in terms of what your country will want in ten years, and we are using the resources available to support the fundamental work that may foresee those wants."

You may say that is selfish, but the sums available through bodies such as ours could be swallowed up if we were to try to achieve the objectives of the ministries.

We are, if you like, being selfish in saying we are being purists. We are supporting work that I believe has a longer lead time than the countries may want, but quite a number of our programs, such as the one on the treatment of sprue, are relevant. Sprue is a a major problem in the South; cerebral malaria, rabies, and snake bite are also relevant. The knowledge of the Belem situation, with its leishmaniasis, is very significant as we watch developments in the Amazon.

I do not pretend we have identified those programs because the ministries said they would like them; that isn't how our limited funds operate. Nevertheless I am totally with you. I believe some organization ought to be doing what the ministries want, because it is they who will put out the fire.

STUART: I did not mean to imply we should be doing what the ministries want. I do believe, however, that the gap between our programs and the concerns of the ministries might not be as wide as we think. Without doing any violence to the objectives of the respective organizations, we may be able to bridge that gap by incorporating them in our planning in a realistic way. And so my plea is that we should seek to bridge that gap.

WILLIAMS: We would be very glad to participate.

LUCAS: It was suggested that TDR has set up high-powered laboratories without paying attention to the rural areas. Of course, Dr. Soprunov, this does not apply to TDR because WHO cannot work without the full knowledge and compliance of the government concerned.

In fact most of our work, especially in the area of training institutions, has been done specifically at the request of governments, to strengthen the research capability of their national institutions.

You know, Dr. Soprunov, that in our dealing with Vietnam, at the Institute of Malariology, Parasitology, and Entomology in Hanoi, that Institution has been strengthened by TDR at the specific request of the government.

In the case of the Americas I could mention the Institute of Investigations in Tropical Medicine in Havana, Cuba, which was set up at the express wish of the government. Another institute in Mozambique was created for the treatment of problems that country considers to be a very high priority. That project is based on rural health, rural epidemiology, and so on.

So, I did not expect Soprunov, as a member of the Joint Coordinating Body, to say the TDR is not dealing with governments, because he knows we are obliged to. We are set up by governments.

WARREN: Soprunov said, "people in rural areas," not ministries of health. Do the ministries always speak for the rural peoples?

LUCAS: In rural areas the problem of onchocerciasis is of concern. In my hometown of Lagos it is not a major problem, but it is the disease of the people in the rural areas. In the most severely affected villages perhaps 15–20% of the people are blind. The TDR program is not intended to benefit the urban areas; it benefits the people in the countryside.

SOPRUNOV: Dr. Lucas, we set up laboratories on the condition that after five years all expenses will be assumed by the government. But if we establish a laboratory in the field of, let us say, molecular biology, that is quite expensive. If that laboratory is to run smoothly and do good work it needs money, and yet the government may not have the funds to invest.

BOOTH: I would like to comment as an historian, an administrator, and a scientist. Obviously, the aim of scientific research in medicine is to benefit people. But whether we are the NIH, the Rockefeller Foundation, the Wellcome Trust, the Medical Research Council, the Soviet government, or the WHO, we are supporting two things that have to be distinguished. The first is scientific research and the advancement of knowledge. The second is application of knowledge for the benefit of people.

Let me first take the developed world, where in any of our countries we find thousands of people suffering from the effects of alcholism, obesity, drug addiction, and smoking. All these are preventive disorders, and theoretically we know all signs, and how to stop people, for example, from having cancer of the lung. If people stopped smoking, they would not get lung cancer.

There are certain situations in which the developed world does function and in which science has produced results that can be lead to enabling legislation. Supposing it were legislated that, for example, motorcycle riders wear helmets. The incidence of head injuries would be reduced by such legislation. One can do the same by insisting that everybody be vaccinated. Vaccination programs have now been established by the WHO throughout the developed world and these prevention programs have had results.

In terms of tropical diseases, we are now talking about vaccines against hepatitis B, and certainly if that develops in science we can go to the health departments of the developed world with that vaccine and ask them to use it.

But please let's separate in our minds that, on the one hand, we are involved with science, which is what Ruth Arnon was referring to in relation to her work in Tel Aviv, and on the other, that the application of science is a totally different ball game that requires a different exercise and a different application.

We know perfectly well, for example, how to prevent malnutrition in the developing world. But are we doing anything about it? The answer is "no."

I think a lot of arguments have to be resolved in terms of scientific discoveries of the same quality as Jenner's and with the development of health departments capable of and effective in applying these discoveries in developing countries.

SELL: I want to thank you for an excellent presentation of the role of the biomedical scientist in the development of knowledge that can be applied. That is the exact position taken by the National Institures of Health. We are in the business of biomedical research, and in fact the government is now considering the creation of a new institute—an Institute of National Health Science Technology, I believe it is to be called—to help in the application of that technology. That is a very important distinction.

There is no question that if we are going to work in this field in other regions of the world with a certain amount of seed money, it may be difficult to continue if that seed money is not followed by the possibility of obtaining funds from other sources.

That is somewhat less of a problem in the United States. The NIH provides funds on an investigator-originated basis. Seventy-five percent of the money the National Institute of Allergy and Infectious Disease puts into all infectious disease research goes to investigator-initiated projects. In fiscal year 1981–1982 less than $10 million is being awarded for grants focused on a particular project. Of that sum, a sizable proportion went to parasitology or tropical medicine.

There is no question that our clientele is submitting grant proposals of the highest quality. The competition is greater than in any of the other 11 institutes, however, and we actually pay a smaller percentage.

Nevertheless, if molecular biological work or any other basic epidemiological work is of sufficient note to convince an investigator's peers, and is favorably reviewed by our study sections, it doesn't matter what disease it is, it will be funded. It is not funded on the basis of program.

GITLER: I tend to disagree with Booth's statement. The only reason we have become involved with the Great Neglected Diseases is because the time is ripe for us to be capable of having an impact in an area in which we don't know the answers.

I was trained as a nutritionist and tried to do something about nutrition in Mexico. I left the field very fast, however, because it was quite obvious that nutrition is a social problem. Today, if I wanted to solve many of the problems being discussed here, I should go out with a placard and cry, "Stop the atomic bombs." But clearly this does not focus on a direct solution to our present topic of discussion.

I agree that we have most of the tools necessary to have an impact on certain diseases that plague us today. I think we can develop the necessary vaccines in a relatively short time and make them available to the developing countries—whether they want to use them or not. If these countries are interested, they have to solve their own problems. There is no doubt about it. They should make use of all the means available to them, including our efforts.

To focus on the problem as I see it: In the past my area of expertise allowed me to enjoy myself doing what I knew how to do best without any concern. But now I have become involved in something I think might have a social impact, and I'm finding myself in a situation where, for the first time in my life, I have no funds to support my own research work.

That is the kind of a vicious circle with which we must concern ourselves. By that I mean, we can have an impact on most diseases we have been discussing. But funds must be available. For example, I can't understand why the TDR doesn't consider amebiasis a disease; to me that is not conceivable. But if we can get funds for amebiasis from another source, that is all right.

If we can't, I have a limited period of productive life so I will have to leave the amebiasis. I have to do the things I think I can do best and where I will have an impact, and this requires funding to follow my ideas and findings. If it happens to be a neglected disease, and other people can benefit, I will be delighted to work on it. But if I have no funds, I cannot contribute to amebiasis and even if committed, I will have to take myself out of the subject.

WARREN: We enticed one of the great macrophage people in the world, Zanvil Cohn, into working on amebiasis, and were able to support him for three years. Cohen then applied to the NIH for a grant and was turned down; he then dropped out of amebiasis research entirely.

Carlos Gitler is one of the great membrane chemists of the world, and he is doing fine work on amebiasis. We can't give him any more support than we are right now, which is not enough, and there is no other source from which he can get funds. It is a truly difficult problem. We do have some sense of responsibility because we have brought many people into the field from others and we alone can't provide enough support. The degree of our support was clear from the outset, however, and it involved a significant amount of funding for an eight-year period. As Peter William said, we are small but somehow we have to make sure the system carries on. That is one purpose of this meeting.

MARTINEZ-PALOMO: I am in a somewhat difficult position because I am the only Third World representative here, and as such I have my own point of view. I believe the future of parasitology lies in the South, not in the North. I mean this because, as has been said here, the North is going to get modern tools in the next ten to 20 years. If we are very lucky we may have new vaccines, but I am somewhat skeptical about that. It is somewhat more likely that we will have improved diagnostic tools. We may achieve a basic understanding of host–parasite interactions, but it is obvious that we are not going to have a healthy world population by the year 2000. I think we all know that.

The basic thing is that the contributions of the new biology to parasitology have to be applied by the South, not by the North. Booth has already commented on this. So one of the basic things to be done, as Lucas mentioned, is to create research and training centers in the Third World.

But attitudes have to be changed, because in general there is still a distinct colonial flavor in the relationships we have. By this I do not mean our relationship with Rockefeller; they have been very generous and we are training people.

But we get scientists from other agencies who are eager to advance their careers, to find something important in a few weeks. They don't care much about the solution of the problem because that takes many years, and they are not willing to spend their lives in a Third World country, trying to solve the health problems, or trying to solve, principally, what has been described as a political problem.

Basically the international agencies have to support research and training centers, but that is not working except on an individual basis. For example, we in Latin America have excellent relations with British scientists; we have practically no contact with Wellcome or with WHO. I would say their support is practically nonexistent in Latin America, which may be partly the fault of our governments. I believe that type of interaction could be greatly strengthened.

One has to be very careful, however, in the sense that it takes generations to establish research centers; they cannot be created in five or ten years. Many countries in the underdeveloped world—in Latin America and in Asia—are not being given support. This has to be taken into consideration, and funds have to go to these centers to strengthen them and give them the opportunity to use the tools of modern biology.

CROSS: I have been very much impressed by the wide diversity of views we have heard from the people here. From the point of view of somebody who is asking for funds to do research, the fact that there is this plurality of motives, approaches, and attitudes is extremely important. Clearly we all agree we need a greater quantity of sources. We have different agencies that provide support from different points of view, and that seems to give pretty good coverage.

As a basic scientist I have always taken the approach that my job is to be a good scientist and to provide a potential solution; it is not for me to know how to apply it in a practical situation. I understand clearly that whatever I and/or my colleagues do with molecular biology, however, it can come to nothing unless the other side of the problem can be tackled.

I suppose something I feel personally at the moment, in relation to the move I am about to make, is that because of the way the Medical Research Council operates in England, and because of my situation at Wellcome, throughout my whole career I have enjoyed stability of funding. That has allowed me, first, to do the things I have done and, second, to attract people into the field, just as some of you here have done. Warren, for example, has done a fantastic job of bringing people into parasitiology.

But it is embarrassing when one finds that there are a great many exceptional and highly motivated young people who want to come into the field, and yet one cannot promise them the kind of opportunity whereby they can tackle a very difficult subject, develop an effective approach, and have the stability of funding to carry on.

DAVIS: The whole kaleidoscope picture of parasitic diseases has to be seen as a symptom in itself of socioeconomic deprivation and ecological, human, and behavioral determinism. We will certainly produce water to damp down the fires of human sickness, of human morbidity, and even of transmission patterns, but unless the developing countries are helped to lift themselves up—and not by their bootstraps—these symptomatic trends of socioeconomic deprivation throughout the Third World will stay with us for a long time.

But these are peripheral to the major problems of human overpopulation, misuse of land, false use of water resources, etc. I believe we will do good, but the time scale has to be measured in many, many decades. Anyone who thinks we can eradicate parasitic diseases is mistaken; anyone who thinks we can control them has to consider a time scale of fifty to a hundred years.

PERLMANN: Three years ago we were not at all interested in anything having to do with parasitology or tropical medicine. Since then Warren convinced us and helped us to become interested in it. We have two laboratories with 20–25 investigators, some from developing countries. We have also established a working contact with two laboratories in developing countries. The GND Program initiated all this.

I personally am still not concerned about parasitology or tropical medicine, but I'm quite aware of the enormity of the problems of these diseases and what they mean.

One of our objectives from the very beginning has of course been to apply our specialized knowledge in immunology toward working on these diseases, where immunology is very important. In addition, I subscribe totally to the pluralistic approach that has been taken. It not only involves the professional background of scientists, but allows different research groups working in this area to have different objectives, and I think that is important.

As I said, we now have 20–25 people in our group, which of course creates a problem. That is the responsibility of what is going to happen to these mostly young people who were enthusiastic and wanted to enter the field. How can we see to it that they will go on? The shortage of funding is the problem now of course. It has been very important for us that the Rockefeller grant was for an eight-year period. In our country, at least, that is unusual; grants are usually for three years, and one cannot start a program such as ours from scratch on a three-year basis.

OPPERDOES: I would like to draw attention to one of the last possibilities left for people working on parasites and in the field of modern biology generally. There is quite a lot of funding by organizations, especially in Western Europe, that used to support basic science. Because of public opinion in Western countries, they are now more likely to give money to projects in applied science. This is an ideal situation for people who work on parasites, which can be used as a modern organism—trypanosomes, for example. When we have problems in getting grants we approach these organizations because now they probably will support a research program that has practical application.

WORBOYS: I would like to make three points. The first is that all the historical evidence suggests that what Davis said about socioeconomic development is exactly right, not just in developed countries, but in former colonies; what we heard this morning about Russia and China confirms that.

The second point is that from what I have heard over the last couple of days, parasitology is now back in the position it held before World War I; it is using the same techniques of virology and bacteriology, but one ought to be optimistic about its relationship to biology, particularly to the new biology.

The third point goes back to Booth's remarks about research and its application. One of the problems with tropical medicine, given its early success with the attack on the vector, is that research and application have always been very closely linked. This has worked both ways for the discipline, and it has advantages and disadvantages.

The experiences in different countries with various kinds of funding suggest that there is no single answer to any of these problems. It could change very rapidly over time, given different political commitments.

BOOTH: May I make an historical point? I agree that research and application were closely linked in the colonial era, when investigators were doing practical work and were responsible for its application. The problem at present is that the research is being done in the North and being applied in the South, and there is no satisfactory link.

9

Research in Parasitology:
Martinsinovski Institute, USSR

Fiodor F. Soprunov

The only way to understand the structure of the Martinsinovski Institute and the work going on in this Institute is to look at it from an historical point of view—Romanovski, Pavlovski. The Institute was established just after the Revolution with the intention of carrying out the eradication of malaria in the country. It was not an easy target. At that time, 50–60 years ago, there were millions of new cases of malaria each year.

The structure of the Institute was adapted to the necessities. A large department of malariology was formed, with many sections—morphology and physiology of the parasite, chemiotherapy and pharamacology, including toxicology and pharmacokinetics, and clinical and chemical synthetic laboratories for development of drugs and insecticides. The second large department was the department of epidemiology and entomology, with a staff of 60 co-workers, including some 30–40 scientists.

At the same time, institutes of malariology have been appearing in the republics of Central Asia and in the Caucasus. Even more important was the creation of a network of malaria stations throughout the country and the building of two factories for the production of the necessary quantity of drugs and insecticides (Paris-green, then DDT).

Many thousands of people were engaged in this work, planned and coordinated in the south of the country.

The number of malaria cases decreased rapidly; there was a temporary increase during the war due to migrations of the population, but after 60 years there were fewer than 100 new cases each year for a population of over 200 million.

Of course malaria was not the only problem in parasitology. Several— helminthiasis, amebiasis, leishmaniasis—were widespread in the country. But these diseases are mainly connected to a low level of sanitary culture and decline and disappear when the level of life improves and the habits of the population change. Last year 20 million coproscopic examinations gave 1.5% positive results for ascaridiasis.

So it happened that after 60 years there was a loss of interest in our country for parasitology and parasitic diseases: no more malaria and a rapid decrease in helminthiasis. The network of malaria stations disappeared, some institutes of parasitology were closed, the industrial production of antimalarial drugs and the search for new drugs were stopped.

A renewal of parasitic work in our country began after 70 years. There were three main reasons for it:

1. Large semi-desert regions were included in the rapid economic development of the country. In west Siberia, and in the central Asian area of Russia, rich natural resources of gas and oil were discovered. The population of these republics is increasing rapidly and new territories are needed for agriculture, especially for cotton production. Ten pumping stations have been constructed and an artificial river is now running upwards for irrigation of the Steppes. You may have an idea of the area of this territory if you compare it with the area of France.

   But in west Siberia you have natural foci of opisthorchiasis and encephalitis, and in the Steppes of Central Asia, foci of cutaneous leishmaniasis. These parasitic diseases became of national importance.
2. Our country is now open to the world, especially to the developing countries in the Third World. Many of our specialists are now working in Africa, Asia, and South American and the importation of resistant malaria cases increased drastically in the last decade. The vectors are now resistant to insecticides and something must be done to avoid the appearance of foci of malaria in the south of the country.
3. The third reason is connected with obligations, our commitment to WHO and the TDR program. We are as a rule critically minded to this program but anyway we realize that something must be done to help the developing countries in the field of tropical medicine. And we all are short of time.

What has been done during the last three years? The structure of the Martinsinovski Institute has been changed so as to deal with the new tools: structure based on taxonomy, structure based on technology and methods.

National program for research and control of opisthorchiasis—less than
    100 research centers
Leishmaniasis 72,000
Malaria

We consider that there are three preliminary conditions to start any eradication program:

1. You must be sure the work can be achieved in one or two decades.
2. You have the necessary money for it, and the people of the region are actively interested in your work and help you.
3. You know that a stable situation will be reached and the disease will not reappear when you stop the eradication work.

For some parasitic diseases we do not plan an eradication in the future.

To conclude I would underline that the development of parasitiology in the USSR is not dictated by traditions of the schools, by imminent scientists, or by financial considerations but from outside by the practical problems that the society has to solve in its development. For the next five years two directions are decisive: (1) realization of three large-scale programs in the country; (2) careful preparation of the starting positions for campaigns of eradication of parasitic diseases in developing countries. Last February (1982) there was a meeting of parasitologists and specialists in tropical medicine of ten countries and the first agreement of collaboartion has been signed; (3) theoretical work for abroad (new components) including specialization of the five institutes we now have—leishmaniasis, biological methods, host–parasite interplay, and socioeconomic aspects of helminthiasis—and a new program in tropical medicine in the framework of the Cameroons.

In conclusion, the structure and the directions of research are not dictated by the traditions of the school or by prominent scientists but only from outside by the problems and difficulties the society has to overcome in its development. As a consequence, we are more interested in people working effectively than in geniuses making new discoveries. We are poor, so we start a program only if the goal that must be reached has been prepared from the beginning till the end, and only if the program is short enough and not too expensive.

10

Parasitological Research in Institutes in China

Mao Shou-Pai

In China, records of malaria (not the parasites) could be traced way back to 300 B.C., and descriptions of *Ascaris, Taenia, Oxyuris,* and *Fasciolopsis* were found in literatures of 700 A.D. Institutional research made its debut in 1928; it is certainly still in its childhood and needs to be carefully nurtured, not to be neglected, and most of all, not to be spoiled.

Research Institutes Before the Founding of the People's Republic

The first institute of parasitology in China, the Institute of Tropical Medicine, was inaugurated in August 1928 in Hangzhou, Zhejiang Province.[1] Dr. S. L. Hung, the originator of the Hamburg flotation technique for hookworm egg counting, was the director of my institute in the early 1950s. It opened with nine technical members to carry out etiological and pathological studies on, as well as control planning against, fasciolopsiasis in Xiaoshan, schistosomiasis in Jiaxing, and malaria in Hangxian. Hookworm surveys were also conducted along the Shanghai–Hangzhou and Shanghai-Nanjing railways. They were the first in China to inoculate laboratory animals with metacercariae of *F. buski* with success and also the first to attempt the cultivation of Hansen's bacillus, the latter without success. The institute experienced ups and downs and became the first provincial institute of parasitic diseases established after Liberation, February 1949.

Two months later, another institute was founded in Beijing under the auspices of the China Foundation for Education and Culture, in memory of J. S. Fan, hence the name, Fan Memorial Institute of Biology.[1] It was mainly concerned with the collection, description, and classification of zoological fauna and botanical flora in China. There were departments of zoology and botany, a museum in Beijing, and a botanic garden in Lushan, Jiangxi Province. Parasitology was involved rather indirectly since flies and snails were studied from a taxonomic point of view rather than as intermediate hosts or carriers of parasites. Nevertheless, it should be pointed out that Chinese pioneers in parasitology, C. Ho and H. F. Hsu, started their careers in that institute. It was closed during the Japanese occupation, reopened in 1946, and merged into Academia Sinica after Liberation.

In 1934, following a flood disaster in China and with the contributions from foreign countries (mainly, if not solely, from the United States), the Central Field Health Station was established under the National Economy Council.[2] Parasitology was one of the nine divisions of the Station, with its objective to study the distribution and transmission of parasitic diseases, and to design a control program. An epidemiological survey and a pilot control project of kala-azar were undertaken in the north of Jiangsu Province with a schistosomiasis survey in Zhejiang Province. These were also interrupted by the Japanese invasion in 1937. The parasitology group moved with the government to the interior and contributed to malaria control in southwest China, especially in the China–Burma border region where strategic communication was vital for military reinforcements. The malaria control program was supported financially and technically by the Rockefeller Foundation.

In 1942, the Central Field Health Station was reorganized to form the National Institutes of Health under the Health Administration in Chongqing. The Department of Parasitology, under the Institute of Epidemiology and Institute of Malariology, was still financially and technically supported by the Rockefeller Foundation.

The endemicity and distribution of schistosomiasis in Sichuan Province were clarified by the Department of Parasitology with a mass survey of parasitic infections among ten thousand laborers participating in the construction of an airfield near Chengdu. A pilot project of malaria control in the steel works in Chongqing City on the Upper Yangzi has contributed much to the understanding of malaria epidemiology and control in that area.

In 1946, both the Department of Parasitology and the Institute of Malariology moved with the National Institutes of Health from Chongqing to Nanjing. In 1947, a Station for schistosomiasis control was established in North Jiangsu under the Health Administration. It transferred to South Jiangsu the following year owing to political instability in the north. Institutional research activities in parasitology were greatly curtailed or abandoned on the eve of the general collapse of the old Kuomintang regime. I remember that I was refused funds for a project costing an equivalent of only 200 U.S. dollars to carry out an experiment on snail control in a newly discovered schistosomiasis focus in the vicinity of Nanjing.

The Department of Parasitology in the Lester Institute for Medical

Research in Shanghai had been very active in the 1930s. Its studies centered on schistosomiasis and insect-borne parasites such as malaria and filariasis. Members of the Department contributed much to the understanding of animal reservoirs, wild as well as domestic, of schistosomiasis. They demonstrated for the first time the existence of schistosomiasis in Yunnan Province. Malaria and filariasis were studied chiefly by entomologists with the view of elucidating life cycle and transmission. The Institute suffered a great deal after Pearl Harbor and could not resume its normal activities until the Japanese surrender. Senior staff members of the Institute became key personnel in different institutions after Liberation.

It would be unfair not to mention the Department of Parasitology in the Peking Union Medical College, PUMC, even though it was committed to education as well as research. Important studies in parasitology included schistosomiasis, clonorchiasis, fasciolopsiasis, visceral leichmaniasis, amebiasis, and hookworm. Three special supplements of the Chinese medical journal devoted to parasitology, pathology, and bacteriology landmarked the contributions of PUMC scientists in parasitology in the 1930s.

In summary, one may say that from 1928 to 1949, research institutes of parasitology in China were few in number and such research was only in its infancy. Nevertheless, it was closely connected with practical needs and the few parasitologists trained during that period became the nucleus of parasitological activities after Liberation, both in educational institutions and in research institutes.

Research Institutes After the Founding of the People's Republic

Institutes

In 1950, the former National Institutes of Health were moved to Beijing except for the Department of Parasitology and the Institute of Malariology, which formed the Huadong (East China) Branch Institute to meet the needs of parasitic disease control in the south. During the first three years, the main effort of this Branch Institute was devoted to the training of teaching personnel for medical colleges which were emerging in the provinces but without an adequate teaching staff. In 1953, the Hainan Malaria Research Station was organized by the National Institutes of Health with most of the senior staff members from the Huadong Branch. In 1956, the former PUMC was amalgamated with the National Institutes of Health to form the Chinese Academy of Medical Sciences. In the following year, the Department of Parasitology at PUMC merged with the Huadong Branch to form the Institute of Parasitic Diseases with Professor L. C. Feng as its director until his death in 1972. The staff of the Hainan Malaria Research

TABLE 10-1. Staff members in the Institute of Parasitic Diseases, CAMS (Dec. 1981).

Scientific (college graduates)	109
Technical	115
Others (librarian, interpreter)	5
Subtotal	229
Administrative	25
Supporting	58
Subtotal	83
Grand total	312

TABLE 10-2. Provincial institutes and their target disease(s).

Institute	Target disease(s)*				
	S	M	K	F	A
Anhui Inst. Schistosomiasis	+				
Fujian Inst. Parasitic Dis.	+	+		+	+
Gansu Inst. Endemic Dis.†					
Guangdong Inst. Parasitic Dis.	+	+		+	
Guangxi Inst. Parasitic Dis.	+	+		+	+
Guizhou Inst. Parasitic Dis.		+		+	+
Hubei Inst. Parasitic Dis.	+	+		+	+
Hunan Inst. Parasitic Dis.	+				
Jiangsu Inst. Parasitic Dis.	+	+		+	
Jiangxi Inst. Parasitic Dis.	+	+		+	+
Shangdong Inst. Parasitic Dis.		+	+	+	
Shanghai Inst. Schistosomiasis	+				
Sichuan Inst. Parasitic Dis.	+	+		+	+
Tali prefectural Inst. Schistosomiasis	+				
Yunnan Inst. Malaria		+			
Zhejiang Inst. Parasitic Dis.	+			+	+

*Abbreviations used: S—schistosomiasis, M—malaria, K—kala-azar, F—filariasis, A—ancylostomiasis.
†Mainly concerned with endemic diseases other than parasites.

Station returned to the Institute but maintained a close collaboration with the Hainan Prefectural Health Station.

Because it is an organization at the national level, the main tasks of the Institute of Parasitic Diseases, Chinese Academy of Medical Sciences, consist of research on technical problems encountered in the control of the most important parasitic diseases. It also trains scientific and teaching personnel for provincial institutes and medical colleges and publishes scientific and technical information. The number of staff members in different categories is listed in Table 10-1.

To cope with the need for disease control, provincial institutes were established in the early 1950s in all the provinces where one or more parasitic diseases were endemic. A list of these provincial institutes and their target diseases is shown in Table 10-2.

Another category of institutes with parasitological studies are devoted to other disciplines but may have some relationship to parasitic diseases. This varies with the interests of the scientific staff. Table 10-3 cites examples of such institutions.

Last, but not the least, are the institutes affiliated with hospitals or medical colleges such as the Institute of Tropical Medicine in the Beijing Friendship Hospital and the Institute of Immunology in Shanghai No. 2 Medical College. In addition, some departments in educational institutions may have an interest in one or another aspects of parasitological research, but they are not research institutes in the strict sense of the word.

TABLE 10-3. Institutes involved in parasitological research.

Institute	Subject
Chinese Academy of Medical Sciences	
Inst. Basic Medical Sciences	Immunology of malaria
Chinese Academy of Traditional Medicine	
Inst. Herbal Medicine	Antimalarial herbs
Chinese Academy of Agricultural Sciences	
Inst. Schistosomiasis	Schistosomiasis in domestic animals
Academia Sinica	
Inst. Zoology (Beijing)	Taxonomy of Mollusca and Crustacea
Shanghai Inst. Materia Medica	Antimalarial compounds
Shanghai Inst. Entomology	Biology and control of mosquitoes
Beijing, Shanghai and Chengdu Inst. Biological Products	*In vitro* cultivation of, and immunization against *Plasmodium*

TABLE 10-4. Classification of 40 Research Subjects in the Institute of Parasitic Diseases, CAMS (1981)

Nature of research subjects	Number	Percent
Applied research on new tools or new approaches	27	67.5
Mechanism studies	12	30.0
Project for wider application	1	2.5
	40	100.0

Research

Generally speaking, institutes affiliated with Academic Sinica are more concerned with basic research, while those involved in health sciences are more oriented to applied research. The provincial institutes are responsible for technical guidance in the control of target diseases in the area. In the Institute of Parasitic Diseases, CAMS, the majority of the research subjects have direct bearing on the control, diagnosis, and/or treatment of parasitic diseases (Table 10-4).

The departmental structure for conducting relevant research varies greatly from one institute to another. In the case of the Institute of Parasitic Diseases, CAMS, at the beginning it was organized in the conventional way into departments of protozoology, helminthology, and medical entomology. When the National Agricultural Development Program (draft) was announced in 1955 to the public by the Central Committee of the Communist government, schistosomiasis, malaria, filariasis, kala-azar, and ancylostomiasis were considered of leading importance. This was based on an article in the draft program ordering Party committees at different levels to eradicate these five parasitic diseases whenever and wherever possible. In 1957, my institute in Shanghai was reorganized and departments were set up relating to diseases. Due to the priority and expertise available, there were six departments for schistosomiasis, that is, epidemiology, diagnosis, biochemistry, pharmaceutical chemistry, pharmacology, and clinics, and one department each for malaria, filariasis, kala-azar, and ancylostomiasis. In other words, the basis for departmental structure was dualistic, by disease on one hand and by discipline on the other. In 1978, with the increasing demand to develop immunologic techniques and chemotherapeutic agents for diseases other than schistosomiasis, departments were finally set up by disciplines, as shown in Table 10-5.

In line with the multidisciplinary research approach, scientists were recruited from various faculties. Table 10-6 shows the career structure of 102 scientific workers in the institute. Very few faculties in the People's Republic of China, include parasitology and medical entomology, while immunoparasitology is practically nonexistent. Usually we find it necessary to offer supplementary training to adapt new recruits to the research demands of our institute.

TABLE 10-5. Departmental Structure in the Institute of Parasitic Diseases, CAMS

Department	Target diseases					
	S	M	K	F	A	O*
Epidemiology	+	+	+	+		
Parasite biology	+	+		+	+	+
Vector biology and control	+	+	+	+		
Diagnosis and immunology	+	+	+			+
Biochemistry	+	+				
Pharmaceutical chemistry	+	+				
Pharmacology	+	+		+	+	
Clinics	+			+		+
Scientific information	+	+	+	+	+	+

*Abbreviation used: O—other parasitic diseases.

TABLE 10-6. Career Structure of 102 Scientific Workers in the Institute of Parasitic Diseases, CAMS

Faculty	Number	Percent
Medicine	37	36.3
Biology*	30	29.4
Public Health	11	10.8
Chemistry	11	10.8
Pharmacy	10	9.8
Others	3	2.9
	102	100.0

*Including microbiology or entomology.

TABLE 10-7. Departmental Structure of 16 Provincial Institutes

By discipline	2
By disease	4
By both discipline and disease	5
By tasks (prevention, clinics, etc.)	5
	16

The departmental structure of provincial institutes is divided into three different categories: by disease, by discipline, or according to actual tasks (Table 10-7). Some institutes may have specialized hospitals for treatment and clinical research on schistosomiasis patients, while others are engaged only in field work and laboratory tests.

Coordination

Adminstratively, there is no linkage between the Institute of Parasitic Diseases, CAMS, and provincial institutes, which are entirely under the direction of local health bureaus, a few of which are under the direction of their local Council of Science and Technology. Communication and coordination are made possible through the National Research Committees; one is for schistosomiasis and another for malaria, with a Research Group for filariasis. The Committees and the Group usually meet annually, with frequent meetings of specific sections between the annual sessions.

Publication

It is rather unfortunate that usually all the papers resulted from parasitological research in the P.R.C. are in Chinese. Most Chinese journals have the titles of leading articles translated into English, while only a few provide an English summary. The name of institutes publishing annual reports is listed in Table 10-8.

As a WHO Collaborating Center, the Institute of Parasitic Diseases, CAMS, has taken over the responsibility since 1980 to prepare English abstracts of all the papers pertaining to parasitology published in Chinese in our journals. We send them quarterly to related headquarters: malaria papers to Malaria Action Program, papers on other parasitic diseases to Parasitic Diseases Program and those on insects and snails to Vector Biology and Control in WHO headquarters in Geneva. They are subsequently reprinted and distributed to interested readers outside China, who may not have easy access to Chinese journals.

I sincerely hope that this brief introduction of parasitological research in Chinese institutes gives some idea of the situation in the People's Republic of China, which had been cut off from the outside world for more than a quarter of a century. Chinese parasitologists are now eager to cooperate with their colleagues in other countries who care to nurture our rising generation without neglecting or spoiling it.

I am grateful to the Rockefeller Foundation for giving me this opportunity to present a brief introduction to the parasitological research in

TABLE 10-8. List of Institutes Publishing Annual Reports

Institute Parasitic Diseases, CAMS
Hunan Inst. Parasitic Diseases
Jiangsu Inst. Parasitic Diseases
Shangdong Inst. Parasitic Diseases
Shanghai Inst. Schistosomiasis
Sichuan Inst. Parasitic Diseases
Yunnan Inst. Malaria
Zhejiang Inst. Parasitic Diseases

Chinese institutes. I wish to acknowledge the valuable assistance of Prof. Yang Xinshie, Mr. Zhao Xing-lie and Mr. He Kai-zeng for providing the references needed. My information would certainly be incomplete without the comments from Dr. Bowers who knows so much about the medical history of China.

References

1. Chyne, WY (1934) *Handbook of Cultural Institutions in China,* Interim Committee, Chinese Association for Global Cultural Cooperation (in Chinese).
2. Tongren Society (1935) *Summary of Medical Affairs in the Republic of China* (in Chinese).

Discussion

BOWERS: Thank you, Dr. Mao. I would like to begin by saying that Mao graduated in medicine from Aurora University Medical School in Shanghai, which was created by the French Jesuits just before World War I. It was believed that one reason they did so was that close by there was Tungchi Medical College, a rather strong German school. The Jesuits thought they should set up their own medical school in a competitive relationship.

Second, the Lester Institute was built by an Englishman, Henry Lester, who was an architect. He made a fortune, and, as I recall, at the time of his death he left an endowment to build the Lester Institute for Medical Research, which was directed by an English physician named Earle, who had been professor of physiology at Hong Kong University.

The head of parmacology was not a pharmacologist but a pharmacist, Bernard Reed. He moved from the Peking Union Medical College to the Lester Institute and did excellent research on the Chinese materia medica.

The Peking Union Medical College was created by the Rockefeller Foundation and dedicated in 1921. It was the "Johns Hopkins of China." It had classes of 25 students, maximum, and copied Johns Hopkins almost assiduously.

It is important to know that it had an excellent field program in parasitology led by Ernest Faust with Henry E. Meleney. Their field studies in schistosomiasis japonicum, published in 1921–1922, reported that approximately 200 million Chinese were exposed to that disease in the Yangtze flood basin.

Dr. Mao, is any of the work on materia medica, which was so active in China, approaching the schistosomiasis problem today?

MAO: We are working on chemical synthesis, and find that the extract of Qing Haosu, which is active against malaria, is also active to some degree against schistosomiasis, especially the synthetic extract. But the most exciting thing about materia medica in China right now is the development of an antimalarial drug, which seems to be a reasonably good one and has a totally different structure than any other antimalarial drug produced. It may be a significant addition to the armamentarium against malaria.

WARREN: That is an important development.

MAO: Dr. Bowers mentioned Drs. Earle and Reed, and I want also to mention Dr. Robinson, who was head of the Department of Parasitology. It was he who discovered schistosomiasis in Yunnan province. It was unfortunate that during the Japanese occupation Robinson was imprisoned in Hong Kong, where he committed suicide.

LUCAS: It is interesting to learn about the background of the research now going on in China. Dr. Mao of course has difficulty in reconciling honesty with modesty. He has not praised the current work of scientists in China. One group you mentioned has studied traditional antimalarial drugs, and made a number of derivatives, some of which are more active and have other desirable features. Two of these derivatives are now going through toxicological studies and rapid development, in collaboration with the Special Program, because they show features such as being effective against resistant strains of *P. falciparum;* they also act much faster than chloroquine. One of the derivatives is being considered as a potential drug for the treatment of cerebral malaria in areas where the parasites are resistant to other drugs.

Most of this work has been done by Chinese scientists. We have collaborated with them and, through training grants, we hope to enable them to extend this work.

The work done so far in London and elsewhere has amply confirmed the reports received from Chinese scientists on the antimalarial properties of these compounds. It is expected that in the next two years of so it may be possible to carry out more extensive trials of the drug. It is not the only one, however; Dr. Mao's units are working on other antimalarial drugs and we are collaborating with them. They are also working on other parasitic diseases (filariasis and schistosomiasis).

11

The Australian School
of Parasitology:
Current Status and Prospects

Graham F. Mitchell

The proceedings of a previous conference on parasitology in the United States,[1] and the other contributions in this volume on the more global picture, contain a wealth of considered opinion of what the discipline of parasitology and its various subspecialties are all about. Because of this it may be pertinent for me to focus on a topic in which I can relay some firsthand information, that is, provide an update on the Australian School of Parasitology.[2] I hasten to emphasize that this analysis is through the eyes of a relative newcomer to the field. On the threshold of the biotechnology era, it may be useful to examine parasitology, and essentially research in parasitology, in this one corner of the globe. Are Australian parasitologists well placed to contribute to the projected revolution in biomedicine and to the continued strengthening of the discipline of parasitology in terms of both the accumulation of fundamental biological knowledge and practical outputs of research? Will we be able to attract the brightest and best of the medical, veterinary, and science graduates into this expanding field of research?

The parasitology scene in this country is not so large as to preclude even an attempt to summarize current research programmes and to make some generalizations and projections. However, there already exist an impressive number and type of institution involved in parasitology research in this country (Table 11-1). If an Australian parasitologist is defined as someone who sometimes, if not regularly, attends the annual meetings of the Australian Society of Parasitology, then the group size is relatively small (350 members). It has been quite noticeable over the past three years or so that national microbiology, biochemistry, and immunology meetings now have "up-front" sessions on parasitology (principally the rapidly evolving fields of immunoparasitology and molecular parasitology) rather than the parasitologically oriented investigators being accommodated

TABLE 11-1. Parasitology research in Australia—locations

Institutes
  Institute of Medical and Veterinary Science, IMVS (Adelaide, South Australia)
  John Curtin School of Medical Research (Canberra, Australian Capital
    Territory)
  Kolling Institute of Medical Research (Sydney, New South Wales)
  Queensland Institute of Medical Research, QIMR (Brisbane, Queensland)
  Waite Agricultural Research Institute (Adelaide, South Australia)
  The Walter and Eliza Hall Institute of Medical Research (Melbourne, Victoria)

Federal Government
  Commonwealth Institute of Health (Sydney, New South Wales)
  1st Malaria Research Unit, Royal Australian Army Medical Corps (Ingleburn,
    N.S.W.)
  Commonwealth Scientific and Industrial Research Organization, CSIRO, Divi-
    sions of Entomology, Wildlife, Agricultural Research and Soils; Division of
    Animal Health: McMaster Laboratory, Sydney, N.S.W.; Pastoral Research Labo-
    ratory Armidale, N.S.W.; Long Pocket Laboratory, Indooroopilly, Qld.; Park-
    ville Laboratory, Melbourne, Victoria.

State Governments
  Queensland Department of Primary Industries, DPI
    Animal Research Institute (ARI), Yeerongpilly, Qld.; Tick Fever Research
      Centre, Wacol, Qld.
  N.S.W. Department of Agriculture
    Veterinary Research Station, Glenfield, N.S.W.; Chemical Research Institute,
      Rydalmere, N.S.W.
  Victorian Department of Agriculture
    Attwood Veterinary Laboratories, Westmeadows, Victoria
  Regional Veterinary Laboratories of Victorian, South Australian and Western
    Australian Departments of Agriculture

University Departments, Hospitals
  Department of Biochemistry, University of New South Wales (Sydney, N.S.W.)
  Department of Clinical Microbiology, Flinders University (Adelaide, S.A.)
  Gastroenterology Unit, Prince of Wales Hospital (Sydney, N.S.W.)
  Department of Haematology, Concord Hospital (Sydney, N.S.W.)
  Departments of Medicine and Micobiology (Princess Margaret Children's Med-
    ical Research Foundation), University of Western Australia (Perth, W.A.)
  Department of Parasitology, University of Queensland (Brisbane, Qld.)
  Department of Paraclinical Sciences, University of Melbourne Veterinary Clini-
    cal Centre (Werribee, Victoria)
  School of Veterinary Science, Murdock University (Murdoch, W.A.)
  Department of Tropical Veterinary Science, James Cook University of North
    Queensland (Townsville, Qld.)
  Department of Veterinary Pathology, University of Sydney (Sydney, N.S.W.)
  Departments of Zoology, Plant Physiology, Microbiology and Paediatrics (The
    Adelaide Children's Hospital), University of Adelaide (Adelaide, S.A.)
  Department of Zoology, Australian National University (Canberra, A.C.T.)
  Department of Zoology, University of New England (Armidale, N.S.W.)

TABLE 11-1. (Continued)

Museums
  The Queensland Museum
  The South Australian Museum

Commercial Operations
  For example, Ciba-Geigy, I.C.I.

somewhere outback. It is a fact that many basic immunologists in this country are rediscovering, in a sense, the field in which immunology was born and in which its greatest triumphs can be found—the immunology of infectious diseases. Host–parasite interactions provide fertile ground for biological research, and involvement of "non-parasitologists" in the analysis of parasites and the intricacies of host–parasite relationships can be expected to accelerate. For example, much information will be provided in the near future by immunologists using metazoan parasites to study allergic phenomena and the regulation of IgE antibody production. The expertise of trained parasitologists is becoming critical in guiding laboratory-based, fundamental biomedical research workers along useful research pathways. The latter individuals, and particularly the molecular parasitologists, will lose perspective and purpose unless very close linkages are formed with persons familiar with parasites, the diseases they cause, and the precise needs for practical new methods of control. Of course, more biological/epidemiological expertise is also required to assess whether or not a new reagent is useful under conditions prevailing in "real life."

On the medical side, Australia currently has no parasitic disease which is of *major* public health importance although the zoonosis echnococcosis (hydatid disease) is a public health problem in particular locations. Filariasis and malaria disappeared from northern Australia as a result of surveillance, prompt treatment of patients, and attention to vector breeding sites. One suspects that intestinal protozoa and nematodes such as *Giardia lamblia, Strongyloides stercoralis,* and the hookworms are more of a problem than currently realized, particularly in Aboriginal communities. The zoonoses toxoplasmosis and toxocariasis have unknown prevalence. In this part of the world where the north–south relationships are reversed, the main foci of human parasitic infection and disease are to the north, that is, Papua New Guinea, the Philippines, Indonesia, other parts of Southeast Asia, and Oceania. Parasitology receives short shrift in the curriculum of Australian medical schools.

Contrasting with the situation of human medicine, parasites of veterinary importance are abundant, and because of the economic importance of the plentiful parasites of sheep and cattle in this country, Australian investigators have been at the forefront of veterinary parasitology. Veterinary research on parasites in Australia has been dominated for years by epidemiological studies of high quality, and the formulation and testing of control measures based largely on managerial practices and appropriately timed administration of chemicals. Parasites involved have been, in the main, the gastrointestinal nematodes and liver fluke of sheep and cattle, tick-borne protozoal diseases of cattle, and ectoparasitic arthropods of sheep and cattle.

"Preventive medicine" has been the key term on the Australian veterinary scene for many years, and my information is that veterinary parasitologists were instrumental in highlighting the cost-effectiveness (and basic common sense) of this approach. To a veterinary student in the early to mid 1960s the movement away from the rather more dramatic therapeutic/first-aid approach to more studied preventive medicine was clearly evident. The status of preventive medicine in the teaching curriculum can be gauged from the contribution of Drs. Arundel and Rickard[3] in this volume on the teaching of veterinary parasitology, at least in the University of Melbourne. For example, analysis of when pasture contamination by parasite eggs can be reduced markedly by judicious application of anthelmintics and exploitation of natural environmentally based attrition of contaminating parasites on pastures (i.e., anthelmintic administration at times of the year most inimical to survival of free-living stages) must lead to profound reductions in subsequent exposure to parasites in susceptible young sheep and cattle. Emphasis in the teaching of veterinary parasitology is on practical aspects, this being right and proper bearing in mind the fact that the bulk of graduates is destined for practice or government departments involved with the rural industries. The farmer is naturally more interested in what the consultant veterinarian knows about practical means to deal with a parasitic disease problem, and the likely cost, than in whether the vet knows the latest on recombinant DNA technologies.

Parasitology has a long history in Australia, being influenced as in Europe and the United States by schools of zoology.[1] The reader is referred to excellent reviews on the early days by Sprent[2] and Mackerras.[4] The greats of the distant and more immediate past include the Bancrofts, Mackerras, Cleland, Breinl, Gilruth, Johnston, Roberts, Clunies-Ross, Fairley, Sprent, Rogers and Gordon, many working in Queensland.[5,6] During my enforced reading excursion into history, it was a source of some pleasure to learn of the magnitude of the Australian contribution, and I am indebted to Ken Warren and the organizers of the Bellagio conference for this enlightenment!

As mentioned, the strength of parasitology research in Australia for a quarter of a century has been in the parasitology of grazing livestock under extensive farming practices. More particularly, the study of parasite population dynamics, the formulation of effective control measures based on chemotherapy/chemoprophylaxis plus pasture and animal management, and analysis of factors contributing to, and means to combat, the development of resistance to chemicals in helminth, acarine, and blowfly populations. The appearance of resistance to chemotherapeutic agents in nematodes, flukes, and ectoparasites has been rapid and spectacular in Australia and the monitoring of this is currently highly developed.

What do Australian parasitologists see as Australia's unique contributions to, or highlights of, parasitology research? Or even "firsts" if there is ever such a thing in global endeavours. The question of Australia's special contribution has been put to about 30 parasitologists, or at least investigators using parasites for biological research (see Acknowledgments). The following is a composite of the responses (with some editorial additions); principal investigators are indicated, the work having been performed in Australia (i.e., excluding work in Papua New Guinea, etc.).

1. Epidemiology (parasite population dynamics) of gastrointestinal nematodes of ruminants and the principles and application of integrated chemical and managerial measures to control nematodes in sheep and cattle grazing extensive areas (Gordon).
2. Discovery of the adult filarial worm, *Wuchereria bancrofti* (J. Bancroft).
3. Large-scale application of nematode parasites for the control of insect pests, in particular the sirex wood wasp (Bedding and Akhurst).
4. Development of preventive surgery for cutaneous myiasis in sheep— the Mules operation.
5. Pathophysiological (e.g., nutritional) consequences of parasitic infection in sheep (Symons).
6. Chemotherapy of experimental malaria infection and parasite life cycle characteristics in man (Fairley et al. at the Australian Army Land Headquarters Medical Research Unit during World War II).
7. Epidemiology of babesiosis in cattle and development of an attenuated live vaccine (Callow, Mahoney).
8. The biochemistry of nematode growth and moulting (exsheathment) (Rogers, Somerville).
9. Description of the life cycle of various ascaridoids (Sprent).
10. Formulation of theories concerning the evolution of balanced host– parasite relationships (Burnet, Sprent, Dineen).
11. Description of primary amoebic meningo-encephalitis (Fowler and Carter).
12. Description of the "self-cure" phenomenon in sheep (Stewart).
13. Description of tick paralysis in dogs due to *Ixodes holocyclus* (Clunies-Ross).
14. Taxonomy of the parasites of Australian native fauna (Beveridge, Arundel, Spratt, most recently, plus numerous early investigators[2,3]).

15. Epidemiology and control of fascioliasis (Gordon, Boray).
16. Development of epidemiological modelling techniques for fascioliasis (Morris and Meek) and gastrointestinal nematode infections (Donald).
17. Biological control of plant nematode pests (Bird).
18. Epidemiology of toxoplasmosis[5] (Munday and Hartley) and sarcocystosis (Munday and Ford).
19. Analysis of parasite genetics (LeJambre, Dobson) and genetic aspects of host resistance in sheep (Dineen) and mouse (Mitchell).
20. Development of prototype larval cestode vaccines in sheep and cattle (Rickard).
21. Studies on the biology of cestodes (Smyth).
22. Demonstration of the immunodiagnostic potential of antiparasite hybridoma antibodies (Mitchell).
23. Analysis of the pathology of acute babesiosis in cattle (Goodger, Wright, and Mahoney) and acute malaria (Clark).
24. Demonstration of resistance of human ovalocytes to parasitization by *Plasmodium falciparum* (Kidson).
25. Establishment of services for monitoring the development of chemical-resistant field strains of ecto- and endoparasites of ruminants (Queensland Department of Primary Industries).
26. Genetic and biochemical aspects of anthelmintic and acaricide resistance (Bryant, LeJambre, Pritchard).
27. Genetic aspects of fly resistance to chemicals and control by gene manipulation (Whitten).
28. Pharmacokinetics and metabolic effects of anthelmintics in sheep (Pritchard) and cattle.
29. Development of a large animal screen for filaricidal drugs (Copeman).

What thoughts do these same Australian parasitologists have on future research needs? The following is another edited listing:

Parasite Biochemistry and Physiology

1. Comparative biochemistry (and, in particular, intermediary metabolism) of parasites at different life cycle stages.
2. Parasite genetics (parasite and parasite cell hybridization, attenuation, and intraspecies variation). Development of genetically defined strains of parasites.
3. Gene organization in protozoa, helminths, and arthropods using recDNA approaches.
4. Analysis of the spectacular chemical resistance problems in Australia (in both endo- and ectoparasites).
5. Nematode moulting and functions of nematode excretory/secretory products.

6. Tissue tropism of parasites.
7. Development of optimized *in vitro* culture systems.
8. Analysis of nonspecific inhibitors of protozoan parasites (e.g., in serum).
9. Physiological factors which regulate parasite numbers.
10. Analysis of mechanisms, genetics, and reversion to susceptibility of drug resistance plus development of slow-release methods and means to prevent development of drug resistance.

Immunology

1. Development of prophylactic vaccines and improved immunodiagnostic and seroepidemiological procedures (for antigen and antibody).
2. Breeding for increased resistance to parasites and increased responsiveness to vaccination plus identification of reliable markers to be used in selection procedures.
3. Analysis of the interrelationships between the immune system and chemicals (effects of the immune system on the efficacy of chemotherapy).
4. Immunology of ectoparasites (ticks, blowfly strike, etc.).
5. Development of selective immunostimulants.
6. Interrelationships between parasitic infection and allergic disease.
7. Analysis of the immunological aspects of periparturient relaxation in resistance.

Epidemiology/Taxonomy/Zoology

1. Quantitative measures of basic ecological and epidemiological parameters (in areas of differing climatic characteristics and involving long-term measurements) for use in computer simulation models of predictive value.
2. Description of parasites of native fauna including aquatic parasitology.
3. Research on methods of extension (to farmers).
4. Development of "on-property" methods to assess parasite damage (particularly the impact of subclinical infections on production systems in the livestock industries).
5. Development of tools for biochemical/molecular taxonomy.
6. Determination of the prevalence and incidence of parasitic protozoa and helminths in selected Australian populations.
7. Development of integrated control programmes involving exploitation of genetically based host resistance, vaccination, attenuated parasites, management/public health practices, chemicals, etc.
8. Investigation of paramphistomidae in Australia.
9. Determination of the importance of agricultural nematodes and parasites as vectors of virus diseases.
10. Taxonomy of Culicoides and other biting flies.

The Growth of Immunoparasitology

One notable growth area in Australian parasitology, as elsewhere, has been immunoparasitology, the study of the immunological aspects of host–parasite relationships. Principal outputs of research in immunoparasitology are the development of prophylactic (and perhaps therapeutic) vaccines against infection and/or disease and the development of improved immunodiagnostic or seroepidemiological reagents (Table 11-2). Immunoparasitology is a relatively new field, as evidenced by the statement " . . . only in the last 20 years has it been generally accepted that parasites induce immune responses in infected animals and man."[7] That this impression was around must be inconceivable to the immunoparasitologist of the 1980s. Parasite immunochemistry (a sort of bridge between immunoparasitology and molecular parasitology) has taken a dramatic leap forward in recent years, admittedly from a very low basal level.

Very early in the history of what can now be identified as immunoparasitology, Sprent in Queensland formulated one of the conceptual lynchpins of the discipline: the concept of stages in the evolutionary development of balanced host-parasite relationships.[8] This notion, which was an extension of the writings of Burnet on the natural history of infectious disease, emphasised the changing array of parasite antigens which confront the host immune system in *evolutionary* time and the two types of interrelated selection pressures exerted: (1) selection for parasites which do not threaten the life of too many individuals in the host population at least in the prereproductive period, and (2) selection for hosts which limit par-

TABLE 11-2. Research objectives in immunoparasitology

1. Identification of effector cells, molecules, and mechanisms of host-protective immunity to parasites and their target antigens.

2. Identification of immunopathologic host-versus-parasite responses and their target antigens.

3. Development of vaccination strategies for prophylaxis or therapy of parasitic infection or disease.

4. Development of immunodiagnostic and seroepidemiological tests (for detection of antibodies or parasite products) of high sensitivity, specificity, and predictive value.

5. Identification of individuals at risk in terms of the disease manifestations of chronic parasitic infection or aberrant responsiveness to vaccination.

6. Development and application of immunological techniques (immunoassays and antibody probes) for the identification of parasites, analysis of life processes in parasites, and detection of expression of cloned parasite DNA.

TABLE 11-3. Immunoparasitology research in Australia—topics

1. Immune responses, intestinal reactions and genetically based resistance to gastrointestinal nematodes (e.g., *Haemonchus contortus, Nematospiroides dubius, Nippostrongylus brasiliensis, Ostertagio* spp., *Strongyloides* spp.; human, cattle, sheep, mouse, rat, guinea pig; 10 groups, >20 investigators).

2. Immunology and pathology of malaria (e.g., *Plasmodium falciparum, Plasmodium* spp. in mice; human and mouse; 5 groups, ca. 15 investigators).

3. Immunology of cysticercosis/hydatids (e.g., *Echinococcus granulosus, Mesocestoides corti, Taenia hydatigena, T. ovis, T. saginata, T. taeniaeformis;* human, cattle, sheep, mouse, rat; 5 groups, ca. 10 investigators).

4. Immunology and pathology of babesiosis (e.g., *Babesia bovis;* cattle; 3 groups, ca. 10 investigators).

5. Immunology and pathology of onchocerciasis/filariasis (e.g., *Onchocerca gibsoni, Dirofilaria immitis, Litomosoides carinii;* cattle, dog, rat; 3 groups, ca. 5 investigators).

6. Immunology and pathology of fascioliasis (*Fasciola hepatica;* cattle, sheep, mouse and rat; 3 groups, ca. 5 investigators).

7. Immunology of giardiasis, cutaneous myiasis, schistosomiasis japonica, cutaneous leishmaniasis, amoebic meningo-encephalitis, sarcosporidiasis, toxoplasmosis, toxocariasis, paramphistomiasis, rodent trypanosomiasis, and infections with *Boophilus microplus* and *Ixodes holocyclus.*

asite burdens to tolerable levels or which can withstand the deleterious consequences of parasitism.

A listing of current research activities in immunoparasitology in Australia is presented in Table 11-3. Although the immunology of gastrointestinal nematodes is a major research topic, it is a fact that the parasite immunochemistry has barely started. What antigens of resident gastrointestinal nematodes (or trematodes, cestodes, and protozoa) gain access to the host immune system and which of these are potential targets of the multicomponent host-protective immune responses in the gastrointestinal tract? Just as the hybridoma technology has been a shot-in-the-arm to all aspects of immunoparasitology, the analysis of antigens in complex metazoa in particular will benefit enormously from applications of this powerful technology.

The immunology of malaria (No. 2, Table 11-3) in this country really started only with the availability of the Trager and Jensen culture technique for *Plasmodium falciparum.* All Australian groups working on the immunology or pathology of malaria are linked to the Papua New Guinea Institute of Medical Research (Director, M. P. Alpers) in Goroka and Madang, PNG. Analysis of the target protein antigens (in blood-stage PNG

P. falciparum) of *in vitro* inhibitory antibodies from patients, or hybridoma antibodies, and aspects of the molecular biology of *P. falciparum*, are well advanced. The latter studies are being assisted by the availability of various isolates of the bovine intra-erythrocytic protozoan parasite, *Babesia bovis*, which differ in antigenic constitution and virulence. The living attenuated *B. bovis* vaccine for cattle, which has been available in Queensland for many years, is currently being dissected in terms of potential host-protective antigens. No one underestimates the difficulties of developing effective defined-antigen (molecular) vaccines against such infections in which antigenic variation in parasites is likely to occur. However, testing of new vaccines will certainly be easier in cattle than in man.

Substantial progress has been made by Rickard and colleagues on vaccination against cysticercosis and analysis of the immunology of veterinary and laboratory cysticercosis/hydatids. Infections with larval taeniids at least in mouse and rat models are characterized by strong concomitant immunity (syn. premunition, i.e., resistance to reinfection in already infected hosts). The rodent–*T. taeniaeformis* systems also provide rare examples of what may be a relatively simple immunological mechanism responsible for expression of genetically based variations in susceptibility to first infection.[9] From the available information, it is predictable that vaccination against first infection with larval taeniids should be highly effective in even the most genetically susceptible individuals. Vaccines which are effective in the genetically susceptible (i.e., those which are most in need of protection) may be far more difficult to develop in several other systems.[9] Again from what is already known in the larval cestode models, a relatively low level of presensitization (with small amounts of the appropriate antigen, or perhaps even immunization with immunogenic anti-idiotype antibodies) may be sufficient to induce high level resistance to first infection. The incoming parasite, at the time of initial infection in the vaccinee, provides a potent stimulus to the presensitized immune system, which responds rapidly by production of host-protective complement-fixing antibodies. Elimination of the parasite is effected prior to expression by the parasite of parasite-protective, anticomplementary activities. It would be something of a relief for the immunologist if high titered, preexisting host-protective immune responses did not have to be induced by a molecular vaccine (in the appropriate adjuvant).

Investigations concerned with the immunology of filariasis/onchocerciasis are at a very primitive stage, and progress in this field has been slow, principally because of the well-known lack of good animal models for the human diseases. The last of the "big six" (Table 11-3) is fascioliasis and, here again, real progress in the identification of potentially host-protective immune responses has been painfully slow. Analysis of immune evasion mechanisms utilized by *Fasciola hepatica* in its wide range of vertebrate hosts will be a profitable line of investigation. At least a dozen other host–

parasite systems virtually complete the Australian *immuno*parasitology research endeavour (No. 7, Table 11-3).

The Coming Era—Molecular Parasitology

It is quite clear that much of experimental parasitology will soon be dominated by the techniques of molecular biology. Already available are a variety of molecular probes (e.g., hybridoma antibodies and cDNA) for the identification of protozoan parasites, as are probes for the analysis of gene organization and rearrangements in parasite differentiation. Highly sensitive immunoassays for detection of expression (protein production) in various recDNA vectors will also become commonplace in the next few years. The term *molecular parasitology* should not be used in the restricted "molecular biology" sense (i.e., analysis and manipulation of nucleic acids) but rather the analysis of processes and structures of parasites at the molecular level—for example, metabolic pathways, antigens including the neglected glycolipids and carbohydrates, surface recognition structures. The comparative approach will assuredly find wide usage, for example, the comparative metabolic biochemistry, comparative immunochemistry, and comparative nucleic acid chemistry of parasites differing in virulence (infectivity and pathogenicity) as well as life cycle stages (of any one parasite). Parasite heterogeneity (intraspecies variation) will have new meaning in the molecular biology era.

There is currently an urgent need for cloned isolates or at least genetically restricted families of parasites; much of molecular parasitology will founder without genetically restricted parasite isolates (preferably clones) to facilitate interpretation of data. The reductionist approach of analysis of genetically restricted parasites in devising techniques and in the elucidation of principles will be necessary before proceeding to the far more complex field situation in which populations of hosts of different genotypes are infected with populations of parasites of different genotypes. Many deficiencies in our biological knowledge must be filled before the molecular parasitology can proceed optimally. For example, progress in this area will be slow unless *in vitro*-derived parasites in pure form and large numbers are available on a regular basis. Four Australian groups are presently involved in the molecular biology of malaria, babesiosis, leishmaniasis, and/or helminths, viz., The Walter and Eliza Hall Institute of Medical Research, The Queensland Institute of Medical Research, CSIRO Division of Molecular and Cell Biology, and Department of Zoology, Australian National University. The number of Australian parasitologists engaged in the high technology will assuredly increase. Fortunately, there is no "colonial cringe" attitude that because major advances are more likely to be made in countries with far greater resources and manpower we should sit back, let it happen elsewhere, and simply import the products.

Merging the New with the Old in the Australian School of Parasitology

At times, one can perceive a degree of suspicion developing between classical parasitologists and the more epidemiologically oriented investigators on the one hand and the basic, more molecularly oriented investigators (or disciples of the "new biology") on the other. (One can also perceive what seem in 1982 to be rather exaggerated expectations for the new biology in the *short* term.) The problem has no real prominence but should be addressed, as difficulties could emerge if the limited pool of funds available for parasitology research in this country was channelled into the growth areas *at the expense* of the established school. Happily, this is an unlikely event bearing in mind the diverse nature of Australian institutions involved in parasitology research (Table 11-1) and the enlightened view of the need for "polyvalence" in the research effort. It would be a great tragedy if, in the short term, the molecular parasitology effort was seen as some kind of threat which will render redundant all the skills built up over many years in parasitology, and which will be essential for the translation of new information into practice with useful outcomes.

The epidemiologists and classical parasitologists certainly need new tools and more precise assays and probes for the characterization of parasites, the construction of reliable and predictive epidemiological models, and the analysis of disease caused by parasites. Alternatively, the molecular parasitologist will waste time unless there is a strong biological input into the work provided by the more established school. Much is lost by strict confinement of the analysis of parasites in isolation in the laboratory and divorced from the realities of natural host–parasite relationships. These points are so obvious as to be trite. The much-sought-after "breakthrough" obtained in the laboratory is really only identifiable in retrospect: an observation made under the controlled conditions of the lab either stands the test of time or falls when experience proves the observation to have validity and to be useful or shows it to be an aberration.

It would be another great tragedy if the impression existed that unless one was involved with hybridomas and recDNA that one was not really engaged in useful parasitological research. The biotechnology must be grafted on to the existing parasitology research effort and must not intimidate investigators into necessarily changing research direction in a radical manner. No super vaccine, when developed, or new wonder drug will render redundant the need to explore all aspects of host–parasite relationship in precise detail. No safe, effective, quality-controlled, stable, adequately tested, and cheap parasite vaccine is just around the corner.[10] Drugs (chemicals) will continue to be the mainstay of parasitic disease control for some time and will continue to be used in combination with public health measures (in human medicine) and managerial practices (in veterinary medicine). Epidemiological studies must continue to receive high priority. Novel approaches to disease control will only come from thorough, *quantitative* analyses of pathogenicity, immunoregulation, genetically based and environmentally mediated variations in host and parasite, and all other biological, as well as molecular, intricacies of the multitude of highly evolved host–parasite relationships in man and livestock. Model systems will elucidate principles and provide clues on what to look for, or how to interpret observations made in the more complex field or clinical situation. It is essential that the future research effort of the Australian School of Parasitology reflect the spectral nature (the continuum) of any avenue of biomedical research from the poles of the very applied to the very basic with strong laboratory-based research being performed in conjunction with high-quality field work; there is little doubt that it will.

Acknowledgments

I wish to thank the following colleagues who provided letters, often detailed with much time and effort involved, to an enquiry about unique aspects of parasitology in Australia, thoughts on future directions, and the nature of research projects in their own or associated laboratories: N. Anderson, J. H. Arundel, I. Beveridge, A. F. Bird, J. C. Boray, L. L. Callow, D. B. Copeman, R. J. Dalgliesh, J. K. Dineen, V. Duncombe, J. Dunsmore, A. Ferrante, G. E. Ford, D. I. Grove, R. S. Hogarth-Scott, A. Husband, L. F. LeJambre, D. F. Mahoney, D. S. Nelson, W. L. Nicholas, W. F. O'Sullivan, A. D. Parkinson, M. D. Rickard, R. I. Sommerville, J. F. A. Sprent, K. J. Turner, and J. Walker. I also wish to thank Dr. J. H. Arundel for several valuable comments on the manuscript.

References

1. Warren KS, Purcell EF (eds) (1981) The current status and future of parasitology. Josiah Macy Jr. Foundation, New York.
2. Sprent JFA (1966) The Australian School of Parasitology. Aust J Sci 29: 40–44
3. Arundel JH, Rickard MD (This volume) Teaching veterinary parasitology
4. Mackerras IM (1948) The Jackson Lecture: Australia's contribution to our knowledge of insect-borne disease. Med J Aust 1: 157–167
5. Doherty RL (1978) The Bancroft tradition in infectious disease research in Queensland, Parts 1 and 2. Med J Aust 2: 560–563, 591–594

6. Norris KR (1981) Ian Murray Mackerras—1898–1980. Historical records of Australian science. Australian Academy of Science Publication, 5: 98–114
7. Voller A, deSavigny D (1981) Diagnostic serology of tropical parasitic diseases. J Immunol Methods 46: 1–29
8. Sprent JFA (1959) Parasitism, immunity and evolution. In: Leeper GW (ed) The evolution of living organisms. Melbourne University Press, Melbourne, p 149
9. Mitchell GF, Anders RF, Brown GV, Handman E, Roberts-Thomson IC, Chapman CB, Forsyth KP, Kahl LP, Cruise KM (1982) Analysis of infection characteristics and antiparasite immune responses in resistant compared with susceptible hosts. Immunol Rev 61: 137–188
10. Mitchell GF (1982) New trends towards vaccination against parasites. In: Capron A (ed) Clinics in immunology and allergy—immunoparasitology, Vol 2, No 2. Saunders, Eastbourne

12

Biomedical Research at the National Institute of Allergy and Infectious Diseases

Kenneth W. Sell and Karl Western

Five years ago while speaking at the Second Anglo-American Meeting on Tropical Medicine, Dr. Richard Krause, Director of the National Institute of Allergy and Infectious Diseases (NIAID), emphasized the international character of biomedical research. His lecture was based on a quotation from Louis Pasteur, who in 1876 stated, "Science knows no country because knowledge belongs to humanity, and is the torch which illuminates the world." Since that time Dr. Krause has consistently and forcefully supported the international scope of the work of NIAID and particularly that work relating to parasitic diseases.

Four broad objectives provided the basis for this commitment.[1] These include:

1. The strengthening of tropical medicine in U.S. universities within the framework of existing biomedical disciplines, such as internal medicine, pediatrics, pharmacology, biochemistry, and immunology.
2. The extension of current U.S. research to the developing countries through "linkages" between American investigators and those investigators in countries where tropical diseases prevail. This required the development of a new research grant program for "International Collaboration of Infectious Diseases Research" (ICIDR).
3. The establishment or strengthening of centers of excellence in developing countries.
4. Expanding opportunities for research training in the United States for young medical scientists and health professionals from developing countries. This training was to include the initiation of a program for "International Tropical Diseases Research Fellowships" sponsored by NIAID in cooperation with the World Health Organization (WHO) and administered by the Fogarty International Center.

In this paper, I will review four topics which relate to these objectives and to parasitology research.

1. The current status and future prospects for funding of parasitology research.
2. The role of the Laboratory of Parasitic Diseases of NIAID in parasitology research and training.
3. The effects of the "new biology" and its relevance to research in tropical medicine and parasitic diseases.
4. The progress which has been made toward achieving the four objectives outlined just five years ago to accelerate research in parasitological diseases at NIH.

Funding of Biomedical Research in Parasitology

A recent conference, "Current Status and Future of Parasitology," was sponsored by the Rockefeller Foundation and the Josiah Macy, Jr. Foundation in New Orleans in 1980. At this meeting, Dr. Irving P. Delappe of NIAID discussed the funding of grants and contracts in the area of tropical medicine, which were supported by the NIAID in the years 1978, 1979, and 1980.[2] He showed a continual increase in support over this period.

Dr. Joseph A. Cook presented more general data on the sources of funding for training and research in parasitology. His presentation was thorough, and I refer you to his paper for specific information and complete analysis of funding support for parasitology in the United States.[3] The profile for funding has not changed much in the past two years. Table 12-1

TABLE 12-1. Estimated support of parasitological research by United States government agencies, fiscal year 1979 (thousands of dollars)

Agency or Department	Amount
Public Health Service	$16,365
Army	11,584
Agency for International Development	10,200
Agriculture	7,028
Navy	1,995
Centers of Disease Control	1,400
Gorgas Memorial Institute	227
National Science Foundation	33
Total	$48,832

TABLE 12-2. Expenditure for parasitology research in fiscal year 1979 (in thousands of dollars)

	Amount
United States government	$48,832
WHO–World Bank	10,771
Private industry	22,327*
Foundations	5,992

*Incomplete as all industries did not report funding.

reproduces the data presented by Dr. Cook showing that the U.S. government spent close to $50 million on parasitological research with the largest contributor being the Public Health Service, and the largest segment of the Public Health Service being the NIAID. For comparison, we can see in Table 12-2 a summary of all sources of funding for parasitology research that were identified by Dr. Cook in 1979.

Clearly, the support contributed by the U.S. government, private foundations, industry, and WHO all taken together provides a meager total when viewed in relationship to the devastating worldwide effect of these diseases. It is, nevertheless, noteworthy that the major funding for parasitology research continues to come from the United States, even though morbidity or mortality from parastiological disease is relatively uncommon in this country. This commitment apparently represents the conviction of both our scientists and politicians that the United States has a special responsiblity for health in developing countries, a responsibility which unfortunately is not sufficient to generate enthusiasm to support major new targeted initiatives or noncompetitive increases in funding.

The National Institutes of Health (NIH) grew rapidly between 1955 to 1965, when appropriation levels grew 14-fold. This reflected a new Congressional interest in science and a federal commitment to support American science generously with tax dollars. This growth shot up in the early 1970s, propelled politically by the Congressionally mandated war on cancer. The NIH budget has since risen to over $3 billion. The NIH has spent the past year treading water. The response of Congress to the health interests of U.S. citizens in 1970 is shown in Figure 12-1, which demonstrates the vast new resources made available for the study of cancer and heart disease. Currently, over $1 billion is targeted for cancer research alone. In constant dollars, the funds for NIAID remained relatively unchanged through 1975. As shown in Figure 12-2, NIAID has made an improvement in its funding posture since 1975.

NIAID remains the primary source within NIH for funding of research in the field of tropical medicine and parasitology. An analysis of research

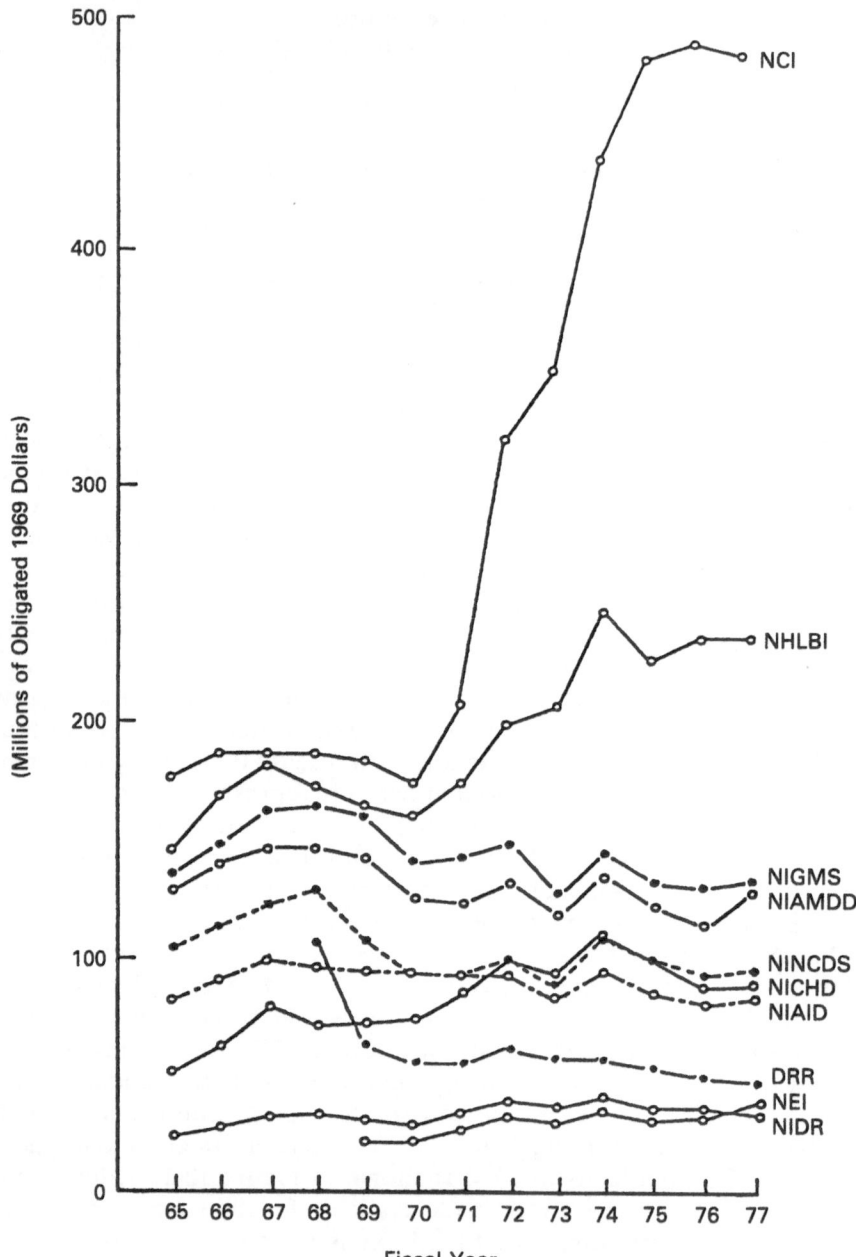

Fig. 12-1 Funding for the B/I/Ds

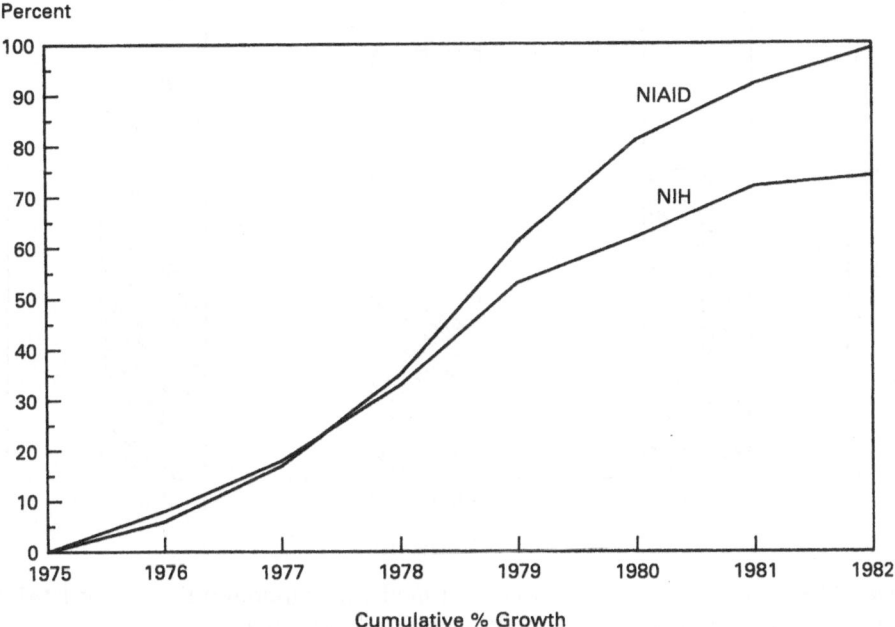

Fig. 12-2 Percent growth in appropriation for NIH and NIAID.

publications in the period 1970 to 1976[4] is shown in Figures 12-3 and 12-4: 30% of published papers in the field of parasitology attributed their support to NIH and of these more than 80% were funded by NIAID.

During Fiscal Year 1981 NIAID continued to assign special priority to its tropical medicine program, particularly to filariasis, leishmaniasis, leprosy, malaria, schistosomiasis, and trypanosamiasas—the six diseases targeted for intensive research by WHO (TDR). The total NIAID commitment to these six tropical diseases was $13.2 million, of which $9.0 million was extramural grants and contracts and $4.2 million intramural research (Table 12-3). Table 12-4 shows the allocation of $4 million for other areas of parasitology research, of which $1.2 million was spent intramurally and $2.8 million used for extramural grants and contracts. In addition we can see that NIAID spends almost $4 million on International Centers for Tropical Medicine Research and Tropical Medicine Research Units, as well as to provide training and career development in the area of parasitology and tropical medical research (Table 12-5).

The International Centers for Medical Research (ICMR) were phased out by NIAID in May of 1980. This program had been in operation since 1960 and had originally been established by the NIH Office of International Research under the authority of the International Research Act of 1960. Responsibility for this program was transferred to NIAID in 1968 and

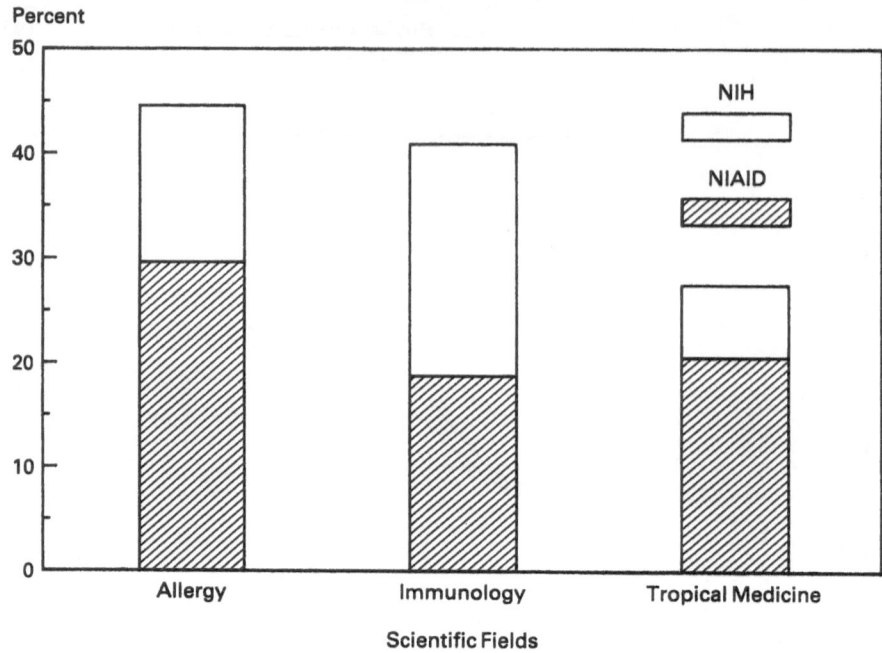

Fig. 12-3 Papers in each area of clinical medicine supported by NIH or NIAID (1970–1976 combined, 275 Biomedical and Related Journals).

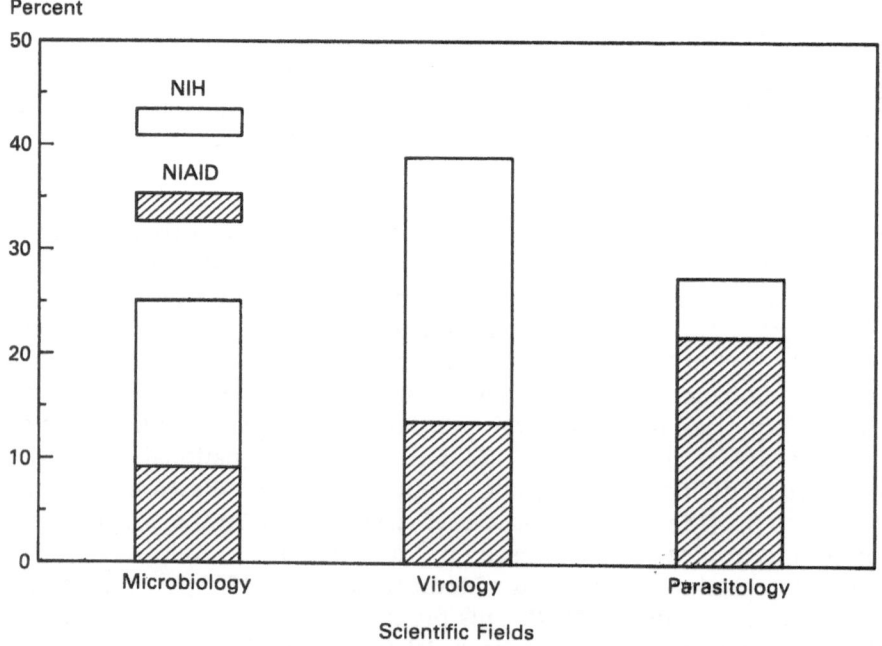

Fig. 12-4 Papers in Biomedical Research Supported by NIAID and NIH (1970–1976 combined, 275 Biomedical and Related Journals).

146

TABLE 12-3. NIAID support for TDR* program

Tropical diseases program area	Grants NO.	Grants AMOUNT	Contracts NO.	Contracts AMOUNT	Intramural NO.	Intramural AMOUNT	Total NO.	Total AMOUNT
Filariasis	11	$ 765,792	3	—	1	$ 411,740	15	$ 1,188,532
Leishmaniasis	13	1,210,950	—	—	1	716,153	14	1,927,103
Leprosy	7	698,928	4	$280,344	—	—	11	979,272
Malaria	17	1,297,787	—	—	5	1,128,011	22	2,425,798
Schistosomiasis	25	1,914,071	1	119,179	3	951,166	29	2,984,416
Trypanosomiasis	31	2,813,941	—	—	4	968,956	35	3,782,897
Total	104	$8,701,469	8	$399,523	14	$4,187,026	126	$13,288,018

*WHO–World Bank Program on Tropical Diseases Research.

TABLE 12-4. NIAID support for general parasitology research

Program	Grants NO.	Grants AMOUNT	Contracts NO.	Contracts AMOUNT	Intramural NO.	Intramural AMOUNT	Total NO.	Total AMOUNT
Cestodes	3	$ 154,806	—	—	—	—	3	$ 154,806
Nematodes	20	1,471,702	—	—	—	—	20	1,471,702
Protozoa	7	453,645	—	—	6	$1,162,536	13	1,616,181
Trematodes	2	177,254	1	$119,179	—	—	3	296,433
Combined	4	422,668	—	—	—	—	4	442,668
Total	36	$2,700,075	1	$119,179	6	$1,162,536	43	$3,981,790

TABLE 12-5. NIAID grants which provide general support to parasitology research

	Number	Amount
ICIDR program	5	$2,307,540
TRU program	2	715,534
Training	17	780,787
Career development	5	158,549
Total		$3,962,410

has continued to provide a stable base for research and training through research centers located overseas. Collectively, tbe four ICMR units— Kuala Lumpur (Malaysia), Dacca (Bangladesh), Lahore (Pakistan), and Cali (Colombia)—served as a national resource to provide a pool of investigators in tropical diseases and to stimulate young scientists to pursue careers in international biomedical research. NIAID has superseded the ICMR program with award mechanisms which are intended to provide greater flexibility and cooperation between U.S. tropical medicine specialists and scientists in developing countries. These mechanisms are (1) International Collaboration for Infectious Disease Research (ICIDR), which now have been extended to ten grants; (2) the Tropical Diseases Research Units (TRU), which now number two; (3) the International Tropical Disease Research Fellowships (ITDRF), currently two grantees; and (4) senior international fellowships which emphasize tropical diseases, which currently has one incumbent.

There is also a new tangible sign of cooperation between U.S. agencies in the support of centers of excellence in underdeveloped countries. A participating agency service agreement (PASA) between the U.S. Agency for International Development (USAID) and NIAID was established to administer a multiple-year project entitled "The Epidemiology and Control of Arthropod Borne Diseases in Egypt and Israel." This project, which began in September 1981, has a first year funding of $1.5 million and will provide for reserach contracts with the Hebrew University in Jerusalem and the Ain Shams University in Cairo. It will also involve participation of U.S. scientists from other federal agencies and universities. Initial emphasis of this program will be the epidemiology and epizoology of Rift Valley fever, malaria, and leishmaniasis.

The total spent in Fiscal Year 1981 is $21,232,218 as compared to Fiscal Year 1980 expenditures of $20,290,000, an increase of about 5% (Table 12-6). For completeness, I should point out that NIH also spends an additional $7.2 million in the general area of tropical medicine including cholera and leprosy, mycology, arbovirology, and vector pathogens, as well as support of ten training grants in these specific areas of interest (Table 12-

TABLE 12-6. Total expenditures for parasitology research

	Fiscal Year 1980	Fiscal Year 1981
TDR		$13,288,018
Other parasitology		3,981,790
General support		3,962,410
Total	$20,290,000	$21,232,218

TABLE 12-7. NIAID expenditures for tropical medicine research other than parasitology

Program	Grant NO.	AMOUNT	Contract NO.	AMOUNT	Intramural NO.	AMOUNT	Total NO.	AMOUNT
Bacteriology	18	$1,608,639	2	$197,343	1	$303,079	21	$2,109,061
Arbovirology	18	2,193,171	2	201,311	1	126,283	21	2,520,765
Vector pathogens	25	1,983,601	—	—	1	264,213	26	2,247,814
Training	10	236,274	—	—	—	—	10	236,274
Total	71	$6,021,685	4	$398,654	3	$693,575	78	$7,113,914

7). If these latter areas of general tropical medicine are included, the overall expenditure of NIAID is $30,094,116 for tropical medicine research.

What are the prospects for next year? As shown in Table 12-8, the U.S. Department of Defense has enjoyed a substantial increase in 1982 and has good prospects for 1983. Hopefully, this will provide additional funding for U.S. Army and Navy initiatives in parasitology research. On the other hand, other government agencies have been less well funded. Research in energy, environmental protection, and transportation has remained stable or has actually decreased. Writing in the *New England Journal of Medicine,* Dr. John Iglehart[5] notes, "Although the NIH budget falls considerably short of the prosperity that Reagan proposes to bestow on defense-related research, biomedicine fares reasonably well in the intense struggle for public support. In order to reach that conclusion, however, one must be prepared to concede that times have changed, that the years of NIH's explosive growth are behind it, and that biomedicine research—as impor-

TABLE 12-8. Research and development funding obligation for major departments and agencies of the U.S. government (thousands of dollars)

	1981 Actual	1982 Estimate	1983 Estimate
Defense	$16,494	$20,553	$24,469
NIH	3,332	3,427	3,533
NSF	964	961	1,033
AID	156	160	166
Agriculture	773	807	838

tant as it is to society—must be prepared to line up with a wide variety of other interests to do battle for the public dollar."

It is projected that NIH may have 14% fewer grants in 1983 than in 1982. The number of research trainees are also expected to drop, although at a somewhat slower pace of 8%. But does this mean that support of parasitology research has a bleak or hopeless future? Dr. Kenneth S. Warren in the final paragraph of his summation of the meeting Current Status and Future of Parasitology[6] stated, "Our greatest concern at this point is that funding agencies have not yet realized the opportunities, so we must do everything in our power to increase their awareness, not for the sake of parasitology, alone, but for the health of people throughout the world." I believe that Dr. Warren may be mistaken. I think that many funding agencies and certainly the NIAID and NIH fully understand and realize the opportunities and responsibilities for the conduct of research in parasitology and the significance of this work for the health of the underdeveloped world. Generation of awareness in our agency is not the problem. Rather we find that the quality of proposals competing for the research dollar often are not adequate to assure sustained or increased funding. Our system at NIH requires that we give support to the best research proposals, although we can identify specific areas for special emphasis such as immunoparasitology. If new proposals of high quality, exploiting new, modern, innovative, thoughtful approaches to the investigation of parasitological problems were submitted to the peer review committees of NIAID, they would receive support and funding, even in these times of austerity. It would seem then that the problem may not simply lay in creating awareness among the funding agencies, but rather creating awareness among our finest scientists in the disciplines of biochemistry, immunology, molecular biology, and membrane biophysics. Only then will the field receive the attention and funding support that it needs and deserves. That this has not already happened to a greater extent is somewhat puzzling. As shown in Table 12-9, the majority of parasitologists are located in basic science departments, where they should be able to communicate easily and cooperate widely in these disciplines.

Intramural Research Program of NIAID

The role of the Intramural Research Program of NIAID in the conduct of parasitology research needs to be emphasized. The Intramural program of NIAID consists of more than 300 doctorate-level scientists working in 13 laboratories. Several laboratories engage in research either directly or indirectly related to parasitology. The largest of these laboratories, Laboratory of Parasitic Diseases (LPD), under the direction of Dr. Frank Neva, is devoted fully to the study of parasitic diseases. The expanding support

TABLE 12-9. Location of parasitologists

	Percent
Undergraduate Programs	
Biology	48.1
Zoology	12.7
	60.8
Basic Science	
Parasitology	8.2
Microbiology	8.6
Pathology	2.1
	18.9
Clinical	
Veterinary	6.0
Tropical Medicine and Public Health	4.0
	10.0

of LPD is no accident. A long-term commitment to international health and biomedical research in parasitology has been consistently maintained within NIAID and is represented currently by over $6.5 million in resources and a total contingent of over 86 personnel (Table 12-10). While the majority of research in LPD is concentrated in the laboratory and clinical research facilities at Bethesda, individual scientists do participate in collaborative studies with scientists overseas, particularly in Brazil. Egypt,

TABLE 12-10. Current staff of the laboratory of parasitic diseases

Staff	Number
Tenured scientists	15
M.D. in training	8
Ph.D.—staff fellows	11
Foreign scientists—training	9
Foreign guest scientists	10
Technicians	26
Secretaries	3
Part time	4
Total	86
In Training	
Domestic	19
Foreign	19

and India. As seen in Table 12-11, LPD has also been a major facility for training of scientists in parasitological research. Over a ten-year period 24 foreign scientists and 46 U.S. scientists entered training in LPD. In addition, a large number of foreign scientists who trained for periods of one to nine months in the laboratory are not included in this breakdown. In Table 12-12 we can see that this pool of trainees provided four tenured scientists who were retained on the staff of LPD. Perhaps more impor-

TABLE 12-11. Scientists trained in the laboratory of parasitic diseases 1970–1980

	Number
Foreign visiting fellows or associates	16
Foreign guest workers	8
Domestic M.D.s—clinical research	27
Domestic Ph.D.s—staff fellows	17
U.S. faculty sabbatical assignment	2
Total	70

TABLE 12-12. Recruitment of tenured scientists from the LPO pool of trainees

	Selected	Total
Domestic trainees	3	44
Foreign trainees	1	16

TABLE 12-13. Fate of scientists trained in the laboratory of parasitic diseases

Still in training	14
LPD	10
Elsewhere	4
In tropical medicine, parasitology, infectious diseases—academic and commercial	38
Ph.D. dropped out of science	3
Physicians who returned to practice	6
Untraced	7
Returned to university faculty	2
Trainee commitment to parasitology research	56/70 (80%)

tantly, as shown in Table 12-13, 80% of the trainees remained committed to research in infectious diseases or parasitology.

The Impact of the Revolution in Biological Technology on Parasitology Research

As with most parasitology research units, the LPD has in recent years changed its focus in order to utilize the modern technologies now available for study of biological problems. For too long biologists have been limited to the study of phenomena, both at the cellular and epidemiological level, with little capability to investigate the precise molecular interactions responsible for virulence, pathogenesis, or protection. This is particularly true in the field of parasitology, where the organisms themselves have special capacities to evade the host protective mechanisms. The parasites can quickly become intracellular, disguise themselves, or present such huge and formidable targets that immune attack is hard to mount.

The modern technologies have made the study of parasites and the interaction between parasite and host much more tenable. Monoclonal hybridoma antibodies can identity specific antigens and, in particular, those antigens that interact with the host to cause either disease or alternatively to elicit an immune response. Utilizing recombinant DNA technology, it is possible to produce portions of the parasite which contain the antigenic moieties that need to be studied. In this regard, I am reminded of the lack of enthusiasm received when we first presented work done on sporozoite immunization at the Laboratory of Clinical and Experimental Immunology at the Naval Medical Research Institute. This work had been undertaken to confirm the important if not landmark findings of Dr. Ruth Nussenzweig, who showed that immunity to a sporozoite antigen produced complete protective immunity in mouse malaria infections. Dr. Richard Beaudoin and our group had confirmed Dr. Nussenzweig's findings, and the work was extended to show that such immunity also could be induced in human volunteers.[7] We were excited by these findings, and they were presented at an International Meeting of Immunology which was held in Sydney, Australia. While the audience appeared interested, they were—to say the least—skeptical regarding the possibility of human immunization with an antigen requiring production in the mosquito gut. This short-sighted attitude was soon overtaken by the rapid progress of science. As often happens, the technological advances push us beyond the deep furrow of our isolated scientific prejudices. The development of recombinant DNA techniques now makes it possible to produce the sporozoite protective antigen in sufficient quantity and purity so that complete immunity to the sporozoite might be achieved. This is not to say that such an antigen is the only important means by which immunization to malaria might be produced. Other stages of the disease may be equally good targets. It is merely to point out that limitations in our technology should not limit our imagination when interpreting important biological facts regarding host immune interrelationships and the potential for induction of protective mechanisms.

Even the recombinant DNA technologies have now been superseded by more innovative methods. Once the amino acid structure of a protein is known or the DNA sequence of the gene which produces the peptide is determined, it is then possible to use a peptide synthesizer to produce the antigen without recourse to living systems at all. Such chemical antigens have already been produced for hepatitis B virus. Synthetic, viral peptides have been tested *in vivo* in mice and chimpanzees and shown to induce antibodies that are specific for the hepatitis B virus. The protective nature of these antibodies is now being investigated. Surely, this approach may also be applicable to the field of parasitology. This is of particular interest because hepatitis virus cannot be cultured, and yet the new technologies may be used to produce unlimited amounts of antigen. In a similar way, we may be able to circumvent the problems of culture of parasitologic organisms, and yet provide specific antigens for diagnosis, prevention, or treatment of parasitic disease.

Even more innovative is the approach of Dr. Alan Sher, who has made an anti-idiotype antibody against the iodiotypic determinants of a monoclonal antibody specific for the *Trypanosoma rhodesiense* organism. The anti-idiotype when injected elicited the appropriate antibody, which then proved to be protective in mice. In this case, no antigen at all, either synthetic or natural, is necessary to induce immunity.

All of these techniques seem particularly important in view of the conclusions that were drawn at the time of the last general discussion of the future of parasitology. A mere two years ago those summarizing the critical needs for the future of parasitology emphasized the requirement for culture techniques. Although these culture systems continue to be important, the use of hybridoma antibodies, recombinant DNA techniques, synthetic antigen production, and the study of anti-idiotypes all might allow us to obviate or at least minimize the necessity for parasite culture. Of course, all of this may be a fanciful, Buck Rogers view of the current state of biology. I suspect not. Four years ago Dr. Krause summarized a conference on malaria and schistosomiasis research by identifying a mood of impatience which existed in the research community as regards progress on these devastating diseases. That mood of impatience is now accentuated as we learn more of the exciting new tools with which to approach the problems of the parasite. There is little question that new knowledge of the parasite and its vectors will continue to be important areas for research. However, now we have new precise tools to identify the host immune mechanisms and to conceive methods of immunomodulation to produce specific and/ or general protection against parasitic invasion and disease.

Progress Toward Implementation of the NIH Plan for Stimulation of Research on Parasitological Disease

How, then, have we responded to the four broad objectives that were outlined for implementing on the NIH plan to support research in tropical medicine and parasitic diseases? Even with our austere budgets, funding for research in parasitology continues to increase, with a 5% improvement in funding from Fiscal Year 1980 to 1981. We have extended the use of special granting mechanisms (ICIDR) to develop more grants which support collaborative research between U.S. and foreign scientists. These hopefully will develop as new centers of excellence for research in tropical diseases in developing countries. With the assistance of the USAID, another program has now been initiated which provides major support for a center in Egypt. Finally, although the numbers are small, training of foreign scientists in tropical medicine has been possible and the NIH–WHO Fellowship Program has been initiated with three current incumbents. In addition, the Intramural program of the NIAID continues to provide a broad base of support, for both domestic and foreign scientists who wish to extend their interests and talents to research in parasitological diseases.

As might be expected, the implementation of the NIH objectives has been incomplete. The commitment is firm, however, and our progress is steady. Many important advances have been made. We have started to attract some of the best scientific minds to supplement the local group of scientists already committed to the study of parasitic diseases. The future of parasitological research depends on our ability to recruit those who have acquired the new tools of the biological revolution for the study of the fascinating and devastating problems of parasitological diseases. The scientists of the "new biology" must be alerted by groups such as this one to the opportunities that lie awaiting them in the study of the great neglected diseases of the world.

References

1. Krause RM (1978) Science knows no country: The contributions of the National Institutes of Health to tropical medicine research. In: Wood C (ed) Tropical medicine from romance to reality. Academic Press, London, pp 245–253
2. Delappe IP (1981) Research in parasitology: The perspective of the National Institutes of Health. In: Warren KS, Purcell EP (eds) The current status and future of parasitology. Josiah Macy, Jr. Foundation, New York, pp 84–93
3. Cook JA (1981) Sources of funding for training and research in parasitology. In: Warren KS, Purcell EP (eds) The current status and future of parasitology. Josiah Macy, Jr. Foundation, New York, pp 121–135
4. Nairn F, Gee HH (1981) An analysis of research publications supported by NIH

1970–1976, NIH Program Evaluation Report. U.S. Government Printing Office, Washington, DC

5. Iglehart JK (1982) Healthy Policy Report. N Engl J Med 306: 879–884
6. Warren KS (1981) Summary of conference. In: Warren KS, Purcell EP (eds) The current status and future of parasitology. Josiah Macy, Jr. Foundation, New York, pp 265–276
7. Rieckmann KH, Beaudoin RL, Cassells JS, Sell KW (1979) Use of attneuated sporozoites in the immunization of human volunteers against falciparum malaria. Bull WHO 57(suppl. WHO 57(suppl. 1): 261–265

Discussion

MITCHELL: One issue that has come up consistently is the question of the low priority given to parasitic disease problems by governments in the "Northern" countries and budgetary restraints. In the future I think we will find more parasitological research going on under the umbrella of "more food production" and "agriculture" as well as "health." Sir Kenneth, do you see any problems in the fact that people involved in a parasitology program might be looking more toward international agricultural programs for basic funding than toward medical sources?

STUART: Of course I think it is a good thing. There is no conflict in looking to those sources for support. Neither does the answer at present lie in putting all our emphasis on parasitology as such. The problem of health in the developing world is really one of approaches.

What I said about parasitic disorders could be said about any other health problems in the developing world. We need to join forces to look for better approaches, better attitudes, better mechanisms for the identification and solution of health problems, rather than be pressured into making individual pushes in individual directions. All of this implies that the infrastructure is not yet ready for this approach.

MITCHELL: If it is predictable that many governments cannot, within their health programs, do very much about parasitic diseases in the short term, as Davis said, shouldn't we be saying, "Let us try to get activity into the agricultural and food production sections?"

As Warren said, instead of malnutrition causing diseases, it is the other way around, so should we not be taking active steps to try and get at the agricultural sources of funds? As mentioned before, it is predictable that less of the health dollar and more of the agricultural dollar will go to parasitic diseases.

LUCAS: This might be an occasion to exploit the interests in agriculture of a particular government and point out to it the need to control schistosomiasis and other related diseases. But it would be incorrect to assume that planning for health is done with less rigor than planning for agriculture. We must face the fact that in both cases developing countries are having great difficulty in unscrambling priorities, and that they are often subject to the whims and caprices of individuals and pressure groups.

What we need to do in the area of health is to ensure that governments have the capability and the interest to identify their priorities.

In a country where malaria and other diseases represent a major public health problem, we should try to get the government to recognize this and do something about it, not because of our interest in parasitology, but because parasitic diseases are important.

That is why some of us put a lot of emphasis on the training of epidemiologists, even though some of our colleagues regard this as a diversion of emphasis. In some countries of Latin America that don't have a program in Chagas' disease, for example, we should support the government in doing national surveys, to find out the size of the problem; and they can then go on to draw up a national program to control the disease.

NELSON: I would like to comment on what Booth and Warren said about British and Commonwealth support of research. One thing that hasn't been mentioned is the enormous contribution made in the past by the Commonwealth Institute of Entomology of the British Museum, which provided the whole basis for our understanding of the taxonomy and biology of all the vectors of medical importance.

The Commonwealth Institute of Biological Control, which was in Trinidad, is also making an important contribution by using parasites for the control of vectors that are of agricultural importance; it is hoping to move into the field of control of vectors of medical importance by using parasites. The British government provides 40% of the budget of the Commonwealth Institute.

In addition to these Commonwealth contributions, they have produced the finest abstracting service of all the international agricultural literature through their 14 abstracting journals.

OGILVIE: I would like to follow up on what Booth and Stuart said. Since I have served with the Wellcome Trust, all applications of quality that have been well received through peer review have been funded. The problem we have is that although we are well aware of the need to introduce new techniques, a lot of the proposals that come to us simply don't have the quality Warren talked about, that is, the combinations of good biology and strong techniques. That is what in our opinion we very much need. I would totally endorse what Warren said.

13

Teaching Medical Parasitology

G. S. Nelson

It is not the purpose of this paper to say how medical parasitology should be taught but rather to draw attention to the importance of the subject in its own right and the need to give parasitology more emphasis in the curriculum of medical schools in the tropics. To this end it is essential to maintain the Departments of Parasitology and Medical Entomology in Schools of Tropical Medicine in Europe and America and to ensure the continuity of the disciplines to help train the future teachers and research workers in this field of Tropical Medicine.

Medical Parasitology is a vast subject which includes (1) the recognition and differentiation of those features of the parasites which are essential for diagnosis; (2) an understanding of the life cycles as a basis for understanding the pathogenesis of parasitic diseases and, (3) a knowledge of the behaviour of the organisms in the host and the environment so that rational control measures can be designed. In almost all Departments of Zoology, parasitology is given equal status with taxonomy, genetics, embryology, palaeontology, or any of the other specialised fields; it is a very popular subject with students studying for degrees in the biological sciences, and it is given considerable prominence in veterinary faculties, but in most of the medical schools in the more prosperous parts of the world, where parasitic diseases are relatively rare and where the curriculum is already overburdened with other subjects, parasitology is usually relegated to a small part of the course in microbiology or pathology. Unfortunately, this is also true for medical schools in tropical countries where the problems of parasitic and vector-borne diseases are of major public health importance. This neglect of parasitology in the tropics is largely due to the creation in the former colonies of medical schools with curricula and examinations which are more appropriate to Europe, but it is also due to the debasement and disparagement of the subject by administrators and grant-giving bodies who mistakenly regard parasitology per se as old-fashioned and out of touch with recent advances in other scientific disciplines.

Parasitology as a Discipline

The study of parasitic and vector-borne diseases is the raison d'être for tropical medicine as a separate discipline. If it were not for the special problems associated with parasitic and vector-borne diseases, there would be no Schools of Tropical Medicine, no Societies of Tropical Medicine and Hygiene, no UNDP/World Bank/WHO Special Programme for Research and Training in Tropical Diseases, no Tropical Panel of the Wellcome Trust, no Tropical Medicine Research Board of the Medical Research Council, and no Rockefeller Great Neglected Diseases Program, and yet many of these organisations fail to recognise the need for professional parasitologists.

The failure to give adequate status to parasitology has had a demoralising effect on the younger generation of research workers in tropical medicine who are ashamed of being identified as medical parasitologists. They are afraid that they will be shunned by their academic colleagues or by the public, which regards all parasites with abhorrence and all parasitologists as cranks. If you want to succeed as a parasitologist, it is better to hide your main interest in parasites under the guise of studying "Geographic Medicine," "International Health," or even "Human Ecology" or better still hide your true identity altogether and join the immunologists, biochemists, or molecular engineers!

Parasitology is a subject that needs to be studied and taught in its own right and not relegated to a subordinate position. Our predecessors did a great disservice to medical science, and especially to the underprivileged people of the tropical world, by neglecting to create a better image for the subject. The academics who helped to establish the medical schools in the tropics failed to establish a career structure for medical parasitologists and they discouraged medical graduates from specialising in parasitology or tropical medicine. In most of the former British colonial territories, there was no provision for specialist parasitology posts in the University establishments and very few in the Government Health Services. The subject is surely of equal importance with ophthalmology, otolaryngology, urology, neurology, gynaecology, and all the other medical specialties, and yet in most of the teaching hospitals in the tropics you will find specialists in these subjects but only very rarely will you find a specialist in parasitology.

When parasitology is given equal status with other disciplines, as for example in the Schools of Tropical Medicine in England, it often heads the list of students' assessements of the relative merits of the different courses offered, but in most of the medical schools in the tropics, parasitology is taught too early in the course, usually by zoologists who are denied access to patients and even to control programmes. It is therefore not surprising that parasitology is treated by the students as one of the subjects which is to be studied and then forgotten after the examination. I have found this to be the case in medical schools in several countries in Africa, Southeast Asia, and the Middle East where I have been invited as External Examiner, but when I have included in my report a note pointing out the lost opportunities for stimulating the students to take an interest in parasitic diseases in the hospital and in the field, my recommendations have been ignored and I have not been invited back for a second term as Examiner! The Deans of the medical schools are usually clinicians and they regard parasitologists as no more than laboratory assistants. The parasitologists themselves have often been at fault because they have made no effort to use living material in their teaching and have missed the opportunity to involve students in field studies. There are of course several exceptions, and we are all aware of the excellent Departments of Parasitology in the medical schools of China, Thailand, Tanzania, and some of the countries of South America.

A few years ago I gave a paper at a meeting, "Training in Tropical Medicine," which was organised by the Macy Foundation in New Orleans.[1] There I recounted the story about the Inaugural Address of the Dean of one of Africa's most prestigious medical schools, in which he said that his ambition was "to slay the ghost of Patrick Manson that walks the corridors of Mulago Hospital"; to give the Dean his due, he wanted to expand his own subject of cardiology and give more emphasis to degenerative rather than infectious diseases, but he was very successful in killing any interest in tropical medicine. He diminished parasitology to a subject which was taught well, but begrudgingly, by a part-time government entomologist, and the Professor of Medicine, who was interested in the nephrotic syndrome, had to send his blood slides to America for the identification of *Plasmodium malariae!* The situation has not improved, and in the neighbouring country of Kenya, the World Health Organisation is helping to create an M.Sc. course in Medical Parasitology and Medical Entomology for Eastern Africa, not in the medical school where it belongs, but on a separate campus in the Department of Zoology. The same is true in Nigeria, where a similar M.Sc. degree for West Africa has been developed in the University of Jos. Both of these courses will have the advantage of being taught by devoted parasitologists and entomologists, but they will be divorced from the day-to-day medical problems of the hospital and because of the non-medically qualified staff, they will find difficulty in working with the Public Health Authorities in tackling the problems of parasitic and vector-borne diseases in the community.

Why have we failed to give parasitology equal status with other specialties in the medical schools of the tropics? Is it merely a question of tradition and an oversight in planning, or are there other reasons? Why is parasitology given subordinate status? It is certainly true that there is far less money to be made by medical parasitologists than by other specialists, and it is often difficult to survive in many developing countries without a supplement to academic salaries. This is usually given as the main reason for the unpopularity of microbiology and pathology in the tropics, but at least there are established posts in these subjects, whereas in most of the medical schools there are no posts for parasitologists or for specialists in tropical medicine. It is one of our dilemmas that we train specialists in these subjects in our Schools of Tropical Medicine in Europe and America but there is often no career for them when they return to their own country. Only when the universities in the developing countries reassess their teaching in relation to their own needs and not in relation to obtaining degrees, which in the past were assessed by external examiners from Europe or America, will this problem be resolved.

Status in Public Health Services

Unfortunately, the status of parasitology in the Public Health Services in the tropics is usually no better than in the academic centres. Medical administrators who are responsible for establishing posts often fail to understand that parasitology requires specialist study. The Health Services are dominated by generalists who sincerely believe that parasitic diseases can be eliminated by improving primary health care. They believe that the existing tools are adequate to control all the major parasitic diseases and they are unaware that there are still many gaps in our knowledge of the parasites and their vectors and that the application of even existing tools requires a high level of expertise. The failure of the global Malaria Eradication Programme was not due to lack of effort and resources but, in many countries, to the failure of medical administrators to heed the technical advice provided by parasitologists and entomologists on the resistance of the vectors to insecticides and of the parasites to chemotherapy. We are all aware of situations where for many years national control campaigns have continued to use DDT, even though it had been clearly shown that the local mosquitoes were completely resistant to the insecticide.

The role of the medical, veterinary, and agricultural parasitologists in the development of the developing world was reviewed in a Symposium of the British Society of Parasitology, entitled "The Relevance of Parasitology to Human Welfare Today."[2] A similar theme has inspired the Josiah Macy, Jr. Foundation and the Rockefeller Foundation to review "The Current Status and Future of Parasitology."[3] There is an interesting contrast in the reports from these two meetings: the British (and notably Nelson!) emphasised with some pride the enormous contribution that classical parasitologists and medical entomologists had made to the opening up of the tropics. In my comments on the WHO Special Programme I said, with only a little exaggeration and a great deal of jingoistic fervour, that as parasitologists

> we have unravelled the life cycles of all the major parasitic diseases of man and his domestic animals. We have described the pathology and natural history of the diseases they cause. We have helped the pharmaceutical industry to produce drugs which are effective against almost all the parasites that affect man and livestock and with our entomological colleagues we have made a deep study of the vectors and their ecology, as well as devising methods for their control. We have laid the scientific foundation on which the World Health Organisation's Special Programme can build.

It was obvious from this symposium that British parasitologists were inspired by a sense of achievement in the past and optimistic about their future role in enhancing the welfare of mankind. On the other hand, the report of the American meeting gives the impression that parasitologists are archaic creatures who have had their day. They are regarded as anachronisms who have served a useful purpose in the past but who must now give way to molecular biologists and immunologists whose primary purpose will be to exploit parasitic models in the interest of basic science but with much less emphasis on the possible benefits to mankind. Of course I am being unfair to my American friends; there are personally ambitious entrepreneurs and self-effacing altruists on both sides of the Atlantic, but there is a fundamental difference in approach. The disparagement of classical parasitology which is so frequently heard in the United States is now infecting Europe, and this must inevitably lead to the further disintegration of the subject. I am alarmed by the effect that this type of disparagement has on the morale of the young scientists in Departments of Parasitology and Medical Entomology. They no longer have confidence that they can attract financial support for their biological research unless it is identified with biochemistry or immunology and scientists working on the taxonomy, ecology, or epidemiology of parasites are discouraged by those who suggest that they are not in the mainstream of "modern" medical science.

Laboratory workers involved in studies on the physiology of the organisms or on the development of antiparasitic drugs, or on the immune reaction of the host, or the development of immunodiagnostic tests or vaccines all require help from professional parasitologists. At the same time, parasitologists need to be aware of the rapid advances being made by experimentalists in other fields which are providing a deeper understanding of the structure and behaviour of the organisms themselves. Our problem is where to find professional medical parasitologists to replace the small core of experienced teachers who will be needed to produce the next generation of medical parasitologists. Ideally, medical parasitology should be taught by medical graduates who have also had formal training in parasitology and who have had practical experience of parasitic diseases in patients and communities. Academics with this background are extremely rare, and most of the teachers in medical parasitology have been either zoologists with very little clinical experience or clinicians who are amateur zoologists. In Britain there have been several great teachers and eminent parasitologists and entomologists in both categories, but those with medical qualifications are rapidly disappearing. Most of our medically qualified parasitologists have been amateurs with an interest in the natural history of parasites in much the same way as amateur ornithologists have an interest in birds. But these medically qualified parasitologists have seen the problems in the field and they have had an interest not only in the parasites themselves but in their behaviour, so that they could devise methods for limiting parasitic diseases in the community. As a result, they have been good teachers and their research programmes have been related to real needs. Their teaching and research have had a profound influence on the development of the tropics. They were the products of a benign imperialism of the Indian Medical Service and the Colonial Medical Services of Asia and Africa. There are very few of these "dinosaurs" left in the British Schools of Tropical Medicine. We need to find a new recruiting ground or devise schemes for giving our young parasitologists more field experience.

If our Departments of Parasitology and Medical Entomology are to survive in the British Schools of Tropical Medicine, we may have to follow the American example and recruit the next generation of medical parasitologists from the large body of non-medically qualified scientists who are interested in this field of research. There is no lack of highly intelligent and enthusiastic young scientists who want to make a career in medical parasitology and medical entomology. For example, every year more than a hundred British and more than a hundred overseas students with good science degrees apply to either London or Liverpool for places in the postgraduate courses for the M.Sc. degree in Applied Parasitology and Medical Entomology. Regrettably, the best British students are turned away because there are only two training awards available each year, and with the increase in fees for overseas students to around £6000 per annum, we are in danger of losing students from countries where the need is greatest. Unless we can find alternative sources of funds for British students and unless we can persuade our governments to be more generous in providing fellowships for overseas students, we could reach the stage where the Schools of Tropical Medicine might cease to function as teaching centres in medical parasitology and entomology.

This is of more than local importance; the Schools of Tropical Medicine in England are an international resource with an international responsibility to maintain their unique collections of parasites and vectors. Nowhere else in the world is there such a diverse group of living parasites and vectors and nowhere is there such a wealth of human experience of the diseases that they cause or transmit. With the present government financial constraints at home and the diminution of Britain's economic role in the tropics, there is a danger that we may have to abandon one of the really altruistic contributions that our country can make to helping the underprivileged people of the developing world. We are fortunate that we still continue to receive research support from the great medical charities, especially the Wellcome Trust, and also from the World Health Organisation and the Medical Research Council, and we have benefited from a recent magnificent grant from the Wolfson Foundation, but our academic base needs to be more secure. Research is possible only in institutions that can afford to pay for the basic maintenance of the structure and supporting staff and yet many grant-giving bodies refuse to pay for overhead. We are reaching the stage where grants for research may have to be rejected because there are inadequate supporting funds either from government or from our own diminishing assets to pay for simple necessities such as heat and light in the laboratories.

If we are to meet the need for training of specialists in the field of parasitology and medical entomology, it is essential that there be continuity of support for the maintenance of the Schools of Tropical Medicine and Public Health in Europe and America. In Britain this is largely a government responsibility through the system of grants to the universities, and also through the support of the Overseas Development Administration, but there is also a need to attract funds from industry and from the more affluent developing countries. In the present period of economic recession, the Rockefeller Foundation and the Wellcome Trust and other private foundations can play a vital role by providing funds for long-term field projects. This will ensure collaborative studies between European and American scientists and local scientists in areas where parasitic and vector-borne diseases are endemic and as a by-product this will help to produce the future generation of medical parasitologists in both the developed and developing world.

References

1. Nelson GS (1974) Training programs for research and teaching in tropical medicine in the London School of Hygiene and Tropical Medicine. Am J Trop Med Hyg 23: 812–816
2. Taylor AER, Muller R (eds.) (1979) Problems in the identification of parasites and their vectors. Symposia of the British Society of Parasitology, Vol 17. Blackwell Scientific Publications, Oxford
3. Warren KS, Purcell EF (eds.) (1981) The current status and future of parasitology. Report of a Conference sponsored jointly by Rockefeller Foundation and the Josiah Macy, Jr. Foundation. Josiah Macy, Jr. Foundation, New York

Discussion

STUART: Nelson has touched on a critically important point, which is the future role of British schools of tropical medicine in the global fight against parasites. There was a time when the important decisions on health in the developing countries were taken centrally in London. The decision-making points have now been dispensed to a number of independent governmental ministries of health.

It is my view that survival of the schools, and certainly their functions, will be closely linked with the activities and concerns of the ministries of health in the developing world.

Although these schools train most of the people who work in the developing countries, they are not seen as central to the concerns of these governments. When there is a threat against the London or the Liverpool school, for example, voices of support should be heard from each major capital in the developing world, but they are not.

The first step must be somehow to internationalize the activities of the two schools. They should be assisted to become integrated into the training and research programs of the countries whose health professionals they help to train. It is no longer appropriate that their activities should be focused so centrally in Britain. The schools are a valuable resource to the developing world, and it would be a calamity if a formula could not be found for sustaining them in this role.

Part of the challenge is also to design a formula by which the developing countries can themselves become more involved and participate more fully in the activities of the schools, while at the same time allowing them to continue to give the guidance which their expertise over the years has put them in such a unique position to provide.

My hope for the future is that Nigeria, Ghana, and all the other developing countries in which parasitic and other "tropical" diseases have high occurrence rates should become directly involved in the activities of the schools and be represented in a more formal way in their programs.

WILLIAMS: I go along very much with what Sir Kenneth Stuart has said. It is extremely important that the resources of the tropical schools should survive. The world needs an international resource that brings together the parasitological and other knowledge from all over the world in some location, and the places, for historical reasons, happen to be in London and Liverpool.

But I also think we are suffering from a definition of "tropical medicine" that equates it with parasitology to such an extent that it diminishes its potential. It is medicine in the tropics, not tropical medicine, that we should be dealing with—that is what the health ministries and the doctors in the tropics are concerned about. The problem is disease in a population that is malnourished. We must look at this holistically, and parasitologists can play an extremely important part. Our field has become isolated because it has limited itself to exotic diseases, instead of staying in the mainstream of medical development.

BOOTH: Within the University of London, where I was professor for 11 years, there are major problems of resources. There are three major post-graduate activities in London: one is the postgraduate school in Hammersmith; the second is the whole range of postgraduate institutes, including those of cardiology and respiratory disease; the third is the tropical school.

So far as the University of London is concerned, it is a straight question of priorities and money. We cannot do everything, so a choice has to be made.

International universities don't exist, and they never will in my view; universities are national organizations. Research institutes can become international, as is the case with the Rockefeller University.

SOPRUNOV: I gather we are talking about basic parasitology and applied parasitology. What is the status of the teaching programs in the United Kingdom at the present time?

NELSON: We try to keep one step ahead, and we try to keep teaching at a higher level—master's degree and postdoctoral level.

Our greatest problem, and one I'm hoping that the Commonwealth might be able to solve, is that we have marvelous applications from all over the world. We have over a hundred applications from people with first- or second-class honors degrees in the United Kingdom itself. We have a demand for something like 40 places in medical parasitology and medical entomology from first-class students. But we have no training awards. When we are lucky, we get one from the Medical Research Council and one from the British Council. When we are very lucky, we get one from the Wellcome Trust.

With the high fees we are paralyzed at present.

SELL: I have two short comments. The first is that I don't like the definition of parasitology. One should define parasitologists as individuals who have made major contributions to the field using certain tools, primarily certain epidemiological tools. I believe it is very shortsighted to think that if people use a different tool they are no longer parasitologists. I don't like that definition at all.

The second thing is that, in terms of programs for the support of parasitological research, there is no question that the schools in England have been important. There is also no question that there are rather substantial programs elsewhere—and we have one of them. It is one of our laboratories that is not an international center. It is intermingled with 12 other laboratories all performing full-time basic research. That laboratory is flourishing. It is getting the support it needs; it is rapidly moving into using new tools; it is rapidly interacting with the other specialties; and it is sitting cheek by jowl with virologists, bacteriologists, and immunologists.

From the point of view of parasitological research or research on medicine in the tropics, we are in a fortunate position.

STUART: It would seem to me that it is no longer realistic to assume that doctors in the developing world will continue to come to London for training in all aspects of tropical medicine. For this reason, I believe that centers will need to be established in appropriate locations in the developing world to enable more of this training to be undertaken locally. Similar centers for training in fields other than medicine are also envisaged.

Possibilities for setting up a number of Commonwealth higher educational centers in the health field are already being actively discussed. There are already in existence certain centers that tend themselves to this type of development and it is on these centers that we should focus our support. These centers could carry on much of what has been the traditional function of the British schools. It would be a pity, however, if these centers developed as separate units without any linkage to each other. The British schools could help to promote these linkages. Apart from this coordinating role, there are already activities and functions of these schools which it would be uneconomical to duplicate elsewhere. I believe there is a realistic role for the Commonwealth to play in these developments.

SOPRUNOV: In my opinion Nelson's story about the contributions was of particular importance to Asia.

DAVID: I felt a little nostalgia in listening to Nelson's talk. We are interested in life cycles in biology, but apparently some people think that we are giving this up for molecular biology and immunology. I can assure you this is not the case.

I have an example relating to the field of surface biochemistry and insecticides. We are taking a whole year because we want to look at developments involving the use of the electron microscope. If we get someone with an eye to look at some of the organisms, we are not going to lose interest in the biology. They can be looked at with highly sophisticated techniques.

For example, there is the biology of the trypanosomes, and how they make themselves transform. That is also being looked at in another way.

LUCAS: I would like to comment on the point Sir Kenneth made about centers of excellence being created in endemic areas. I have seen some of these in Bangkok and Malaysia, but perhaps there is a time span we have to consider. These centers are not as fully developed as they would like to be. On the other hand, the schools of tropical medicine in Europe and elsewhere, with strong departments of parasitology, may need to adjust their programs to take note of these developments in the endemic countries. I believe the institutions in Britain, Belgium, France, and elsewhere will continue to do work that is of very great importance in this field, although what they do will be different from what they are doing now.

In the developing countries especially, one of the ideas I am trying to push is that there are scientists in these countries dedicated to biochemistry, and one of the most useful applications of biochemistry and bio-chemical research in those countries would be in the parasitic diseases.

One final point is that other resources, such as excellent field research units, are about to disappear and are not being replaced. If we could have a smooth transition, instead of having an institution close down and then having to make the painful effort of building something new after a long gap, we would do very well. That is my view of the discussion about schools of tropical medicine in Europe and in the developing countries.

DAVIS: I congratulate Nelson on his usual excellent presentation. I would say, however, that you tend to agonize too much about our tradi-tional problem in Britain. I believe the French may well be coming up with a series of ideas for research in parasitology that might lead the English to cast a few thoughts in that direction.

I might add that there are major schools of tropical medicine all over Europe; the Germans are by no means inactive, the Italians are coming back, and the Russians are also highly active. And so I believe it is clear that the rest of Europe is really marching on.

BOOTH: It is most encouraging to hear that statement of the position of the Liverpool School of Tropical Medicine. It was Dr. Johnson who said there is nothing to so concentrate a man's mind as the immediate prospect of being hanged.

WILLIAMS: I quite agree with Booth. Isn't it nice to see an evolution taking place, because obviously when Nelson was in New Orleans in 1973, he thought he was going to become a dying species. But he knows now that we want him to come along with us, rather than to get isolated from us. We want your expertise. Don't be afraid to give it to us. It is a bigger world than it was, and one that all of us are a part of. The tragedy is that the schools have held back from joining with the rest of the people who care about this thing that you care about.

14

Teaching Veterinary Parasitology

J. H. Arundel and M. D. Rickard

Before writing this paper we read the excellent article by Weinstein[1] which dealt with the teaching of parasitology in the United States and discussed the need to depart from traditional courses and use an interdisciplinary approach in order to capture the imagination of the young undergraduate. This is easier to implement in general parasitology courses than in a specialised curriculum for training veterinary or medical students. Time constraints and the breadth of important information in the latter determine that parasitology cannot be taught by using only a few examples of the various classes and dealing with them in depth. The dilemma faced by instructors in veterinary parasitology is how to marry the effective teaching of the subject as a useful and practical discipline with all the factual data that goes along with this, while also providing an imaginative insight into the basic biological phenomena of the host–parasite relationship. The biological complexity of a single organism may approach that of its host; where is the line drawn?

In our course we deal with over 350 parasites—helminths, protozoa, and arthropods—that are felt to be important in Australia or are of sufficient importance in foreign countries to warrant some knowledge of them by our students. For each of these parasites we feel that the student should understand its zoological classification, morphology, life cycle, pathogenesis of infection, the reaction by the host, epidemiology, diagnosis, treatment, and control. Parasites that are more important because of their pathogenicity or because they exist within the geographic area covered by our School receive more detailed consideration than those of less importance to us, and many can be dealt with in groups. There is, however, still a vast amount of factual data to be given in the time available and it is important, therefore, to have a clear understanding of the aims of the course if the student is to gain an understanding of the discipline rather than learning, and soon forgetting, a multitude of facts.

We must be constantly aware that our major aim is not to produce parasitologists but to give undergraduates a sufficient understanding of the discipline so that, as graduates, they can diagnose, treat, and control parasitic disease. To satisfy this aim, the student requires a sound basic training in zoological taxonomy so that he has an understanding of the relationship of one parasite to another. It is important that the student know which parasites occur in a particular location in the host and can differentiate them. This does not mean that a student should have sufficient taxonomic skill to identify a nematode without information as to its host and location in the host, although we believe that a student should be able to place any new parasite at least in its Order and Family. Further, he must know the life cycle of each parasite, how the different life cycle stages can harm the host, and how the host reacts; a clear understanding of pathogenesis is essential to help the graduate make accurate diagnoses. A good knowledge of the life cycle also helps the graduate select the appropriate diagnostic technique and understand control procedures.

Basic courses in different schools will differ depending upon the importance of parasitic disease within that country and the interests and expertise of the teachers in that school. In general, and with some notable exceptions, veterinary graduates are to be preferred as teachers as they are better able to give the required applied emphasis and to integrate the basic and applied courses. In Australia we are dependent on agricultural products for a large part of our export earnings, and animals are grazed all year on pasture on large farms. Parasites are therefore important aetiologic agents and parasitology has had more time devoted to it than in many other countries, particularly those where winters are severe and where animals are housed. Under extensive pastoral farming systems the herd rather than an individual animal approach is taken with farm animals and emphasis is placed on the epidemiological approach to control parasitism. A similar approach is taken in the United Kingdom but this has largely been neglected in the United States.

An important decision is where to teach parasitology within the veterinary curriculum and how much emphasis should be placed on each of the basic and applied courses. The practice of teaching parasitic diseases of each host species without a satisfactory basic course, which is seen in some U.S. veterinary schools, makes it difficult, if not impossible, to give students an understanding of the discipline. In our course, all of the aspects listed above, with the exception of epidemiology, treatment, and control, are given in the second professional year of the course. At the same time the students receive instruction in microbiology, immunology, and general pathology and the preclinical subjects—physiology, biochemistry, and animal production. This means that the students commence parasitology before they have a full appreciation of pathological processes. Thus a zoological approach is taken initially to ensure that the students have an understanding of the interrelationships between and within the groups of parasites. Later the emphasis is shifted so that, by the end of the year, the students can approach the subject from the basis of the host, and they know what parasites are in each organ, the pathological processes they cause, how to diagnose infection in the live animal, and how to differentiate the parasites on autopsy.

This host approach is given greater emphasis in the practical classes, which we believe are the most important staff–student contact periods for teaching parasitology. It is disappointing to see in a recent survey of professional veterinary curricula by Burridge[2] that 12 of 21 schools in the United States have less time devoted to practical classes than to lectures. Parasitology is a discipline that will be used routinely by almost all graduates and, as such, should be given in a practical context with time devoted to identification and diagnosis as well as to gaining an understanding of the theoretical aspects. We have two lectures and one three-hour practical class throughout a 26-week teaching year. It is important to have a dynamic approach to teaching because in this way some of the fascination of the host–parasite relationship can be captured by the inquisitive student. He can see and evaluate for himself parasitological principles and apply information gained in studies of other biological disciplines to understanding aspects of the biology of parasitism. This means showing life cycles in action, obtaining and examining parasites from host material, differentiating the various parasites in organs presented for examination, and discussing and practising the diagnostic techniques that can be employed with the living animal.

The following are selected examples of exercises that we use to illustrate parasitological principles and to train students in practical techniques. For convenience they have been arranged to show the principles that they illustrate, but in practice a single experiment can be used to demonstrate several principles.

Class Nematoda

A. Life cycles
  1. Modes of infection
     a. Direct infection by ingestion of eggs. *Ascaris suum* eggs are embryonated and used to infect mice from which migrated larvae are recovered.
     b. Direct infection by ingestion of a free-living larva. Sheep faeces containing *Trichostrongylus colubriformis* eggs are cultured, and the infective L_3 are collected and used to infect a lamb, which is later killed and examined for worms.
     c. Indirect life cycle. The blood of a dog naturally infected with *Dirofilaria immitis* is examines for microfilariae.
     d. Transport host. Mice infected with *A. suum* and *Toxocara canis* eggs are killed at 24 hours, 7 days, and 50 days after infection, and their liver, lungs, and muscle are examined for larval stages.

Larvae of *T. canis* can be found in the muscle of the mouse transport host at 50 days after infection but not in mice infected with
A. suum eggs.

e. Prenatal infection. Examine the small intestines of puppies
killed 1, 7, 14, and 21 days after birth for developmental stages
of *T. canis*.

B. Pathogenesis

1. Migrating larval stages. Rabbits are infected with *A. suum* eggs and
the lungs are examined 7 days after infection. Larvae are recovered
and pulmonary damage assessed.

2. Adult parasites. Three worm-free lambs are infected with either
Haemonchus contortus, Ostertagia circumcincta, or *T. colubrifor-
mis*. The lambs are maintained for 4 weeks after infection during
which the following parameters are measured: faecal examination
for nematode ova, haemoglobin levels, plasma total protein, and
serum pepsinogen. Lambs are killed at 4 weeks post-infection, the
autopsy findings discussed, and total worm counts carried out.

3. Abattoir specimens. Material is obtained from the abattoirs and
knackery to show pathological changes due to parasitic infection.
Museum specimens are also demonstrated.

C. Parasitological techniques

1. Recovery of tissue stages

a. Baermann methods. Larvae are recovered from livers and lungs
of rabbits and mice infected with *A. suum*.

b. Digestion. The large intestine of mice naturally infected with
Syphacia obvelata and *Aspiculuris tetraptera* are digested in
pepsin-HCl and the worms are recovered using a sieve
technique.

2. Faecal examination

a. Qualitative faecal examination. A sodium nitrate flotation
method is used to examine dog and cat faeces, and the eggs present are identified.

b. Quantitative egg counts. The McMaster and paracytometer methods using saturated NaCl flotation are used to examine equine,
bovine, and ovine faeces.

c. Larval culture and principles of identification of infective L_3.
Equine and ovine faeces are cultured, and larvae are collected
and identified using keys.

d. Diagnosis of lungworm infection. The Baermann method is used
to examine cat faeces for *Aelurostrongylus abstrusus* larvae.

3. Total worm counts. Dilution methods are used to obtain an estima-
tion of the number of worms present in the abomasum and small
intestine of lambs, and the worms are then identified.

4. Examination of blood for microfilariae. Blood from a dog with
heartworm is examined using the modified Knott or filter
techniques.

Class Trematoda

Fasciola hepatica is used to demonstrate a trematode life cycle, as well as to illustrate particular aspects of its own biology.

1. Eggs collected from the gall bladders of infected sheep are demonstrated.
2. Fully embryonated eggs are observed hatching on microscope slides.
3. Dissecting microscopes are used to show miracidia attacking and penetrating into *Lymnaea tomentosa* snails.
4. Infected snails are dissected to show rediae and cercariae, and the cercariae can be seen encysting on glass petri dishes.
5. Mature metacercariae are excysted *in vitro* to show active juvenile flukes.
6. Students examine abattoir specimens of livers from infected sheep.

Class Cestoda

Members of the family Taeniidae are used to illustrate cestode life cycles and biology.

1. Eggs collected from adult *Taenia pisiformis* from dogs are examined.
2. *T. pisiformis* eggs are artificially hatched and activated *in vitro*.
3. Rabbits and mice experimentally infected with *T. pisiformis* and *T. taeniaeformis*, respectively, are autopsied and cysticerci recovered from them. Cysticerci of *T. hydatigena* and *T. ovis* are obtained from abattoir specimens.
4. *T. pisiformis, T. taeniaeformis,* and *T. hydatigena* larvae are evaginated *in vitro*.
5. Adult cestodes are fixed, stained, and mounted for morphological examination.

Phylum Protozoa

Flagellates. A culture of *Tritrichomonas fetus* is examined as a hanging drop preparation.

Haemoprotozoa. Mice are infected with *Babesia rodhaini* and *Plasmodium berghei* to demonstrate:

1. Pathogenesis of infection, that is, development of anaemia, formation of blood pigments.
2. Diagnosis of infection using thick and thin blood smears and brain squashes to examine capillaries containing infected red blood cells.

Coccidia. Chickens are infected with *Eimeria tenella, E. necatrix,* and *E. acervulina* and examined 3, 5, and 7 days post-infection.

1. The distribution and characteristics of the lesions produced by the parasites are noted.

2. Microscopic examination of smears of fresh material are used to obtain and identify developmental stages. This is supplemented by examining stained histological sections.
3. Unsporulated and sporulated oocysts are examined.

In addition to these exercises, in all parts of the course—that is, nematodes, cestodes, trematodes, protozoa, and arthropods—students are provided with specimens for morphological examination. Fresh material is provided where possible, and students are encouraged to make their own temporary mounts. Where material is difficult to obtain, or scarce, then permanent, preserved specimens are provided.

This dynamic approach is not new. Although this method is often stated to be too demanding on staff to justify its use and while it takes more time than the presentation of mounted material which can be kept for many years, carefully selected demonstrations using parasites available from research workers in the country means that all organisms do not have to be permanently maintained by the staff and that, with experience, the classes can be prepared quickly.

The applied course may be given as a separate course or be integrated within veterinary medicine. Which is done will, in most cases, depend on the interest and expertise of the staff. If the parasitologists do not have an applied interest, then the applied course will be given by staff in the medicine department, but they rarely know the epidemiology of the various parasites sufficiently well to allow them to give the subject in the modern preventive medicine context. It is important that the interaction of parasite control with animal husbandry practices and economics be emphasised for large animals where the aim is economic control of disease, and where each treatment must be economically justified. In contrast, in small animals parasitic infection is often treated, regardless of cost. In farm animals preventive programs should now be available for all host species, based on the epidemiology of infection, which is the sum of the class of animal infecting the environment, the timing of infection, and the ecology of the free-living stages.

The student should know the drugs available, their spectrum of activity, and their chemical relationships and have a clear understanding of their role in the control of parasitic disease. For example, while drugs are often used to treat animals showing symptoms of disease, their major role should be to prevent contamination in the period most inimical to the free-living stages and so cleanse the pastures and maintain low infection for the remainder of the year. In our course this section is given as a part of veterinary medicine by the parasitologists, and further practical classes are also given to emphasise the diagnosis of parasitic disease.

Postgraduate teaching is offered in most veterinary schools, the field of study usually being determined by expertise of the staff in that institution. Veterinary graduates are difficult to entice into graduate work as the poor salaries paid to graduate students does not compare with that obtained in practice or in government positions, and the demand for an adequate remuneration for a graduate in his or her mid-20s is very great. Fortunately, some of the better graduates have an interest in research and others who work in government departments require graduate training for promotion. The majority of graduate students applying for places, however, are graduates with limited training in parasitology. Many of these are primarily interested in using a particular parasite as a tool in biochemical, immunological, or other studies and cannot really be regarded as parasitologists. It is in their own interests to ensure that they receive a broader training in parasitology as part of their graduate program so that they have a better understanding of their basic tool and a knowledge of other parasites that may interact in their studies.

References

1. Weinstein PI (1981) Teaching parasitology: The current scene. In: Warren KS, Purcell EF (eds) The current status and future of parasitology. Josiah Macy, Jr. Foundation, New York, pp 51–62
2. Burridge MJ (1981) Teaching veterinary parasitology in American universities. In: Warren KS, Purcell EF (eds) The current status and future of parasitology. Josiah Macy, Jr. Foundation, New York, pp 63–68

Discussion

WARREN: That is a problem in all professional training schools, including medical school. The Macy Foundation had a conference on teaching tropical medicine in 1973 at which the issue was raised of how much teaching time is devoted to parasitology in medical schools in the United States. The average was about ten hours of lectures in the medical schools, with one or two laboratories and museum material.

I would like to comment on Nelson's earlier remarks about Makerere University. It was my understanding that parasitology was not in the medical curriculum when Makerere was established. There was a U.S. Army parasitology laboratory in Kampala at the time, and the people working in it told me they had to bootleg parasitology instruction to the students because it wasn't in the curriculum. That is a big problem in professional schools.

LUCAS: In teaching a large class of students who are going to become veterinary surgeons, some will be treating cattle and sheep, and others will take care of pets; very few may be interested in doing research.

Isn't it more likely that we will attract potential researchers and build up their enthusiasm through special courses or electives, and perhaps an additional year or so?

MITCHELL: The core course material is really a specialized curriculum for training the general veterinarian. One would hope to identify during that period the kind of special student who will have expanded by postgraduate work or electives. We need to identify that individual early enough to direct him into another channel.

One has to have two types of courses: a postgraduate course to create the person we know will make a major contribution to research in veterinary parasitology, and a course for the individual who will be concerned with diagnosis, control, and treatment of parasitic diseases of animals as a practicing veterinarian.

David: Veterinary immunology has always been in fashion in England, and in the forefront of research on immunology. Why is the situation so different with veterinary parasitology?

Mitchell: The situation is not different with veterinary parasitology; in fact, veterinary parasitology is very much at the forefront of parasitology, at least in the country I am familiar with, Australia.

Warren: Is there a reason veterinary immunology is so strong in England?

Ogilvie: The Agricultural Research Council, unlike the Medical Research Council, puts 95% of its funds into research; it has only about 5% free money. That may be the reason.

Warren: The question is the teaching of immunology in veterinary school. It is probably rudimentary.

Mitchell: It is in some medical schools.

Williams: May I take up Ogilvie's point? The way the veterinary situation has developed in Britain, where the veterinary schools are training general practitioners rather than specialists, has meant that the Agricultural Research Council provides support for research in its own institutes outside of universities. For the most part the strength of veterinary research in Britain—and there are exceptions—lies in those independent institutes.

This has greatly damaged the universities, however. They have not been able to develop a scientific base because the teachers are not researchers, and therefore they don't tend to attract students to take postgraduate courses. Any student in the veterinary schools who shows an inclination for research and is absolutely first class in quality is likely to transfer to one of the institutes, and then take no further part in trying to change the next generation.

The practical implications of veterinary research are significant in most countries, but in Britain a situation has developed where it is not the universities that are fulfilling the practical requirements of the community because of the emphasis on a non-university institute system.

Warren: I agree. We have a teaching problem with schools of public health. Some believe it was a great mistake for the Rockefeller Foundation to found separate schools of public health, which meant that the medical schools felt it was no longer necessary for them to teach preventive medicine and public health because that was being done by the new schools.

The whole business of shifting responsibilities in fields such as research in veterinary medicine, public health, and so on is important to note.

Williams: Earlier I referred to Makerere not having parasitology in the curriculum. As it was set up, parasitology was a public health subject and it was located with community medicine. Medical schools don't generally have departments of public health. The training in public health which I received at medical school was extremely dull.

Booth: My public health education in Scotland was the most interesting course.

Williams: On the London scene the interest in public health transferred to the developing world. The University of London was the mother to the universities that started in the former British colonies overseas.

Nelson: In the Liverpool School of Tropical Medicine, parasitology is taught to veterinary undergraduates. In fact, all undergraduate teaching of veterinary parasitology is done by our Veterinary Parasitology Department.

We also run a master's degree program in veterinary science that is completely integrated with the master's in medical parasitology and applied entomology. At present we have 12 students working for their master's degree in medical parasitology and seven in veterinary parasitology.

Lucas: I taught public health in Ibadan for some 14 years; it was the only course medical students had to take over a five-year period.

Mitchell: I didn't realize a situation existed where veterinary research was institutionally separated from the teaching of veterinarians, which seems a shame. I am surprised they have done as well as they have because we all know that for teachers to do research is a great way of reminding them of the fickleness of biology and how things change. I am full of admiration for the English having done so well in a situation that sounds deplorable.

Williams: It is like it was in Germany with the Max Planck Institutes.

Booth: Theoretically, it is true if one compares it with the Max Planck experience in Germany between the wars, but it can be overstated. There is no question that veterinary science is very strong in Britain, and if we take the members of the royal societies, for example, the veterinarians will do far better than physicians. They constitute first-class scientists.

I don't know any veterinary schools. The one I have been involved with most is the Bristol school in the west of England. I have examined research Ph.D candidates in my own field, and I am very impressed with the work they are doing.

Mitchell: We are institutionalizing veterinary research to the point where veterinarians are doing good research. I wonder about the teaching of courses by practicing veterinarians. Does the teaching suffer when the research is taken away?

The research should do very well because of the time constraints a good teacher has in providing all the material. What would suffer, I think, in that system would be the training of young veterinarians.

Literature of Parasitology

15

The Status of the Parasitology Literature: Linkages to Modern Biology

Kenneth S. Warren, William Goffman, and Eli Chernin

The first part of these studies was presented at a meeting, "The Current Status and Future of Parasitology," in New Orleans in October 1980. In that paper the parasitology journals of the world were listed, as was the evolution of the major English-language journals in the field, the quantity of papers on parasitology in 19 English-language biomedical journals, and a qualitative analysis of the subject matter published in the four major English-language parasitology journals and three major English-language tropical medicine journals. It was concluded that the parasitology literature as defined quantitatively and qualitatively was deficient in subject matter related to the modern biological disciplines of immunology, biochemistry, and molecular biology.[1]

In the present paper the linkages of the parasitology literature to the modern biological disciplines are examined via citation analysis, and the results compared with the literature of virology and bacteriology. In addition, the content of 16 English-language textbooks of parasitology and tropical medicine was assessed. The latter was done becasue of Paul Weinstein's concerns, expressed at the New Orleans meeting, that parasitology textbooks have

> been dominated by the traditional subject areas of morphology, biology and life cycles, pathogenicity and symptomatology, diagnosis and treatment, and epidemiology and control. . . . Biology as a discipline, however, had undergone a revolutionary change in the past few decades, particularly molecular, cell, and developmental biology, and this

has been accompanied at the organismal, population, and ecological levels by the synthesis of new and far-reaching theoretical concepts.[2]

The textbooks were carefully scrutinized to determine the degree to which they reflect the new biology.

Citation Analysis of Four Major English-Language Parasitology Journals

The *Journal Citation Reports* (JCR), published by the Institute of Scientific Information (ISI), Inc., indexes approximately 1000 biomedical journals. For each journal, it lists all journals cited and the frequency of citation by year; it also lists all journals citing each journal and the frequency of citation. Furthermore, JCR provides an impact factor for each journal, which is defined as the average number of citations per published paper. Using this array of information, the following system of analysis was developed. The citation of a paper y by a paper x implies that relevant information has been conveyed from the author(s) of the paper y to the author(s) of paper x. By analogy, we can say that relevant information has also been conveyed from journal Y, which published paper y, to journal X, which published paper x. Consequently there is a flow of information from cited journals to citing journals.

The percentage of citation of one journal by another, that is, the number of times journal X cited journal Y, divided by the total number of citations by journal X, can be thought of as a measure of the information flow from Y to X since this percentage is an approximation of the probability that journal X will cite journal Y. For a given percentage, the flow can be unidirectional or bidirectional. Flow is unidirectional if the measure of information passage from X to Y or Y to X exceeds the designated percentage but not conversely; it is bidirectional if the measure of information flow from both X to Y and Y to X exceeds the designated percentage. Clearly, if two journals are bidirectionally associated, they are more closely related than if they are unidirectionally associated. Moreover, the greater the level of association, the closer the relationship.

Citation analysis of a set of journals representing a given discipline can thus identify those journals, hence those disciplines, which are associated with the given discipline. Such an analysis was carried out for a selected set of parasitology, virology, and bacteriology journals.

Since our primary interest is in the parasitology literature, the set of journals chosen for analysis consisted of a core of parasitology journals and those journals most closely associated with them. The four major English-language parasitology journals *(Parasitology, Journal of Parasitology, International Journal of Parasitology, Experimental Parasitology)* were selected as the core and all journals directly associated with any of these four journals at the 1%, 2%, 3%, and 4% levels were identified from the 1980 JCR. These thresholds, though arbitrarily chosen, seem to be representative of the scale of association among journals with respect to citation patterns. Clearly, the higher the threshold, the smaller the associated set.

Table 15-1 lists the sets of journals unidirectionally associated with the four core parasitology journals at the 1%, 2%, 3%, and 4% levels. The resulting lists contained 28, 13, 7, and 4 journals, respectively. Those journals bidirectionally associated with the parasitology core are italicized.

Table 15-1 shows that the parasitology journals are unidirectionally associated with a number of high-impact journals of the more modern fields of biology at the 1% level, such as *Journal of Biological Chemistry, Journal of Immunology, Nature, Journal of Experimental Medicine, Science, and Journal of Cell Biology.* These relations rapidly evaporate, however, at higher thresholds so that at the 4% level the parasitology journals are associated only with themselves. Bidirectionally, even at the 1% level, the parasitology journals are associated only with tropical disease, protozoology, and helminthology journals and not with journals of the more modern fields of biology.

Unidirectional associations of the four parasitology journals at the 1%, 2%, 3%, 4% levels for the years 1978 and 1979 were also examined. There was little movement toward greater association with the journals of modern biology by the parasitology journals in the three years from 1978 through 1980. For example, at the 1% level, the 1980 list of 28 journals contained only five journals that did not appear on either the 1978 or 1979 lists. These are *American Journal of Veterinary Research, Clinical and Experimental Immunology, International Archives of Allergy and Immunology,* and *Journal of Invertebrate Pathology and Veterinary Parasitology,* none of which is a high-impact journal.

Tables 15-2 and 15-3 list similar data for three major English-language journals in virology and one in bacteriology. Not only are the virology and bacteriology journals unidirectionally associated with high-impact journals of the more modern fields of biology as high as the 4% level *(Cell, Journal of Molecular Biology, Proceedings of the National Academy of Science, Journal of Biological Chemistry),* but they are bidirectionally associated with some of these journals at the 1% and 2% levels *(Cell, Proceedings of the National Academy of Science, Journal of Molecular Biology).* The average impact factors for journals unidirectionally associated with the four core parasitology journals are 2.91, 3.48, and 4.76 at the 1%, 2%, and 3% levels, respectively, but for the three core virology journals they are 6.23, 6.84, 8.83, and 8.83 at the 1%, 2%, 3%, and 4% levels, and for the *Journal of Bacteriology* they are 4.03, 4.42, 5.18, and 5.76.

Thus, journal citation analysis of the parasitology, virology, and bacteriology literatures reveals that the former is relatively isolated from the

TABLE 15-1. Journals unidirectionally associated with four major English-language parasitology journals at 1%, 2%, 3%, 4% levels (1980)

1% Level	2% Level	3% Level	4% Level
Parasitology (1.95)†	Parasitology	Parasitology	Parasitology
Exp Parasitol (1.49)	Exp Parasitol	Exp Parasitol	Exp Parasitol
J Parasitol (0.66)	J Parasitol	J Parasitol	J Parasitol
Int J Parasitol (0.74)	Int J Parasitol	Int J Parasitol	Int J Parasitol
Am J Trop Med Hyg (1.22)	Am J Trop Med Hyg	J Immunol	
Am J Vet Res (1.01)	Can J Zool	J Protozool	
Ann Trop Med Parasitol (0.77)	Immunology	Nature	
Biochim Biophys Acta (2.86)	J Exp Med		
Bull WHO (1.44)	J Immunol		
Can J Zool (0.84)	J Protozool		
Clin Exp Immunol (2.97)	Nature		
Immunology (2.36)	Trans R Soc Trop Med Hyg		
Infect Immunol (2.66)	Z Parasitenkd		
Int Arch Aller Immunol (1.41)			
J Biol Chem (5.71)			
J Cell Biol (9.74)			
J Exp Med (10.65)			
J Helminthol (0.66)			
J Immunol (6.41)			
J Invertebr Pathol (0.77)			
J Protozool (1.38)			
Methods Enzymol (1.64)			
Nature (6.49)			
P Helm Soc Wash (0.30)			
Science (5.70)			
Trans R Soc Trop Med Hyg (1.33)			
Vet Parasitol (0.78)			
Z Parasitenkd (0.62)			

*Those underlined are bidirectionally associated.
†Numbers in parentheses are impact factors of journals

TABLE 15-2. Journals unidirectionally associated with three major English-language virology journals at 1%, 2%, 3%, 4% levels (1980)

1% Level	2% Level	3% Level	4% Level
J Virol (4.32)†	J Virol	J Virol	J Virol
Virology (3.42)	Virology	Virology	Virol
J Gen Virol (2.05)	J Gen Virol	J Gen Virol	J Gen Virol
Biochim Biophys Acta (2.86)	Cell	Cell	Cell
Cell (14.39)	J Biol Chem	J Mol Biol	J Mol Biol
Eur J Biochem (3.60)	J Mol Biol	Nature	Nature
Infect Immun (2.66)	Nature	Proc Natl Acad Sci	Proc Natl Acad Sci
J Biol chem (5.71)	Proc Natl Acad Sci		
J Exp Med (10.65)			
J Immunol (6.41)			
J Mol Biol (5.68)			
J Natl Cancer Inst (3.02)			
Nature (6.49)			
Proc Natl Acad Sci (8.77)			
Science (5.70)			
Nucleic Acid Res (5.07)			

*Those underlined are bidirectionally associated.
†Numbers in parentheses are impact factors of journals

mainstream of modern biological research. Moreover, there seems to have been little significant movement away from this isolation in the 1978–1980 period.

Analysis of Subject Matter of 16 English-Language Textbooks of Parasitology

One of us (E.C.) perused all of the textbooks shown in Table 15-4. On the basis of these detailed studies, it appears that Weinstein's concerns are corroborated; he notes that "one can no longer be complacent about teaching life cycles and systematics in a purely descriptive manner. It is necessary to interpret these phenomena within the context of contemporary biology." Read's and Whitfield's books make an attempt to deal with

TABLE 15-3. Journals unidirectionally associated with *The Journal of Bacteriology* at 1%, 2%, 3%, 4% levels (1980)

1% Level	2% Level	3% Level	4% Level
J Bacteriol* (2.60)†	J Bacteriol	J Bacteriol	J Bacteriol
Arch Microbiol (2.04)	Biochim Biophys Acta	Biochim Biophys Acta	J Biol Chem
Biochem USA (4.67)	Biochim Biophys Res Commun	J Biol Chem	J Gen Biol
Biochem J (3.19)	Eur J Biochem	J Gen Biol	J Mol Biol
Biochem Biophys Acta (2.86)	J Biol Chem	J Mol Biol	Mol Gen Genet
Biochim Biophys Res Commun (2.96)	J Gen Biol	Mol Gen Genet	Proc Natl Acad Sci
Eur J Biochem (3.60)	J Mol Biol	Proc Natl Acad Sci	
Genetics (2.41)	Mol Gen Genet		
J Biol Chem (5.71)	Nature		
J Gen Biol (1.80)	Proc Natl Acad Sci		
J Mol Biol (5.68)	Science		
Methods Enzymol (1.64)			
Mol Gen Genet (2.88)			
Nature (6.49)			
Proc Natl Acad Sci (8.77)			
Science (5.70)			

*Those underlined are bidirectionally associated.
†Numbers in parentheses are impact factors of journals

these problems, but their deficiency is a marked limitation in "formal parasitology." The time is ripe for a major new textbook combining the best of classical parasitology with the "process" of modern biology.

Conclusion

Our studies of the parasitology literature have revealed a major gap in both the journal and textbook literatures between classical parasitology and modern biology. These deficiencies must be corrected in order to enable the great potential of parasitology to be realized through the rapid application of modern biology by a new generation of scientists.

TABLE 15-4. Major English language textbooks concerned with parasitology

Author	Title	Year	Pages	Comments
Major General Texts				
Belding, D. L.	*Textbook of Parasitology*	1965	1374	Reference text organized along conventional lines. 50/1300 pages on metabolism and immunology.
Cheng, T. C.	*General Parasitology*	1973	965	Exhaustive, and most modern of texts. Chemistry in Chapter 2 (17 pages), Immunity in Chapter 3 (25 pages). Electron micrographs (ems) throughout.
Faust, E. C., Russell, P. F., Jung, R. C.	*Clinical Parasitology*	1970	890	A classical text, occasional reference to metabolism, rare ems.
Schmidt, G. D., Roberts, C. S.	*Foundations of Parasitology*	1981	795	Large introductory text. 5 pages on modern immunology. Little metabolism, some ems.
Intermediate General Texts				
Noble, E. R., Noble, G. A.	*Parasitology*	1976	566	Text for college undergraduates. 55/566 on physiology, immunology, biochemistry.
Olsen, O. W.	*Animal Parasites*	1974	562	General parasitology treated exhaustively. Essentially a reference work. Virtually no modern biology.
Short General Texts				
Markell, E. K., Voge, M.	*Medical Parasitology*	1981	374	Condensed, elementary manual, traditional in treatment.

TABLE 15-4. (*Continued*)

Author	Title	Year	Pages	Comments
Beck, J. W., Davies, J. E.	*Medical Parasitology*	1981	355	Same
Brown, H. W.	*Basic Clinical Parasitology*	1975	355	Same
Sloss, M. W., Kemp, R. L.	*Veterinary Clinical Parasitology*	1978	274	Laboratory diagnosis of veterinary parasitoses.
New Approaches				
Reed, C. P.	*Parasitism and Symbiology*	1970	316	General principles related to a small number of examples.
Whitfield, P. J.	*The Biology of Parasitism*	1979	277	Organization by "process"— nutrition, physiology, immunology, population dynamics, etc., but little emphasis on the parasites themselves.
Tropical Medicine				
Wilcocks, C., Manson Bahr, P. E. C.	*Manson's Tropical Diseases*	1972	1164	Classical text, clinical focus. Discussion of immunology for each major disease averages less than one page.
Hunter, G. W. III, Swartzwelder, J. C., Clyde, D. F.	*Tropical Medicine*	1976	900	Parasitoses subsume ⅛ of text. Conventional treatment.
Woodruff, A. W.	*Medicine in the Tropics*	1974	623	Parasitoses subsume ⅛ of text. Opening chapter— Immunology (26 pages). No other basic science.
Maegraith, B.	*Clinical Tropical Diseases*	1980	621	Parasitoses subsume ⅛ of text. Conventional approach emphasizes care of patients.

References

1. Warren KS (1981) The present status of the parasitology literature. In: Warren KS, Purcell EF (eds) The current Status and Future of Parasitology. Josiah Macy, Jr. Foundation, New York, pp 142–151
2. Weinstein PP (1981) Teaching parasitology: The current scene. In: Warren KS, Purcell EF (eds) The current status and future of parasitology. Josiah Macy, Jr. Foundation, New York, pp. 51–62

Discussion

CROSS: I wonder to what extent we are at the mercy of international publishers. I can think of four books not on your list that are published in England, and I suspect that several of the books on your list are not available in England. To what extent do the distributors play a part in this?

WARREN: These books are on the National Medical Library's list of major textbooks, which is supposed to be an international listing. The crucial question is: Are any of the four books missing from this list general parasitology textbooks, and, if so, are they any better than those listed?

CROSS: Perhaps not. But they are, in my experience, books that serve as a very useful introduction for the nonparasitologist. They are *Biochemistry of Parasitic Protozoa,* W. E. Gutteridge; *Parasitic Protozoa,* J. R. Baker; L. H. Chapell's book, *Physiology of Parasites;* and *Modern Parasitology* by F. E. G. Cox.

WARREN: I don't think the Cox book is available in the United States yet.

NELSON: The Smyth book, *Parasitology,* is not on the list and it is a standard textbook in England.

WARREN: We were confronted with the problem that a lot of books we were looking for are not in print. I gather that Smyth's textbook is in print. *Protozoology* is out of print and we couldn't get a copy. A new edition of Solbit's *Veterinary Parasitology* came out in June 1981, but it wasn't obtainable so I didn't get a chance to see it.

DAVID: Two journals that are more specialized, *Parasitic Immunology* and *Molecular Biochemistry of Parasites,* would be in the 100% category.

WARREN: The analysis of the journals was done in 1979. At that time some of them were just beginning to appear so we couldn't analyze them. They are different, however.

SOPRUNOV: The Japanese are very high on parasitology, and some textbooks have been published in Japanese. You didn't mention the *Southeast Asian Journal of Tropical Medicine Public Health.*

WARREN: That's right; we dealt with a relatively small number of journals. We couldn't examine all of the parasitology journals in terms of subject matter.

SELL: The citation analysis is an extremely interesting way of finding out how journals relate to each other and how scientists relate to each other at the cutting edge of science. We are using that technique in working with the Institute of Scientific Information (ISI) in Philadelphia to analyze papers published by individuals supported by the institute, and to determine who quotes them, and who quotes the people who quote them.

We have been coming up with small groups of people who deal with subtopics in the area. Then, by using time analysis, we determine who publishes first, who is quoting first, and that sort of thing. One begins to get an idea of who relates to whom in the field. One also gets a sense of who is on the leading edge of that field. All this can be done by ISI, which already has in its computer every article published in English identified by subject.

WARREN: The ISI covers about a thousand journals, but not every article; there are 20,000 journals.

SELL: Not in the areas in which we publish. It is a good way, incidentally, to analyze whether or not a certain laboratory is making an impact.

WARREN: If any of you are interested, about two years ago William Goffman did a citation analysis of the six WHO diseases and analyzed the citations in that whole field, with authors, and so on. We could provide that to people who want it.

DAVID: There is a danger, however, because Warren's list doesn't include the modern subject. For example, malaria was not cited, but eosinophils were. One has to be wary about that.

WARREN: I think if we keep moving forward we will get to it, but one can't really be at the true cutting edge of science doing it that way.

SELL: We used the technique to produce our annual report. Instead of asking each laboratory to list what they publish each year, we simply go to the ISI and pull out everything under "publications." As we all know, such listings run anywhere from three to 12 months behind. Nevertheless, it tends to be more thorough than the scientists themselves are.

WILLIAMS: What is your conclusion?

WARREN: The conclusion we have been concerned about all along is that the literature is a mirror of the field. The conclusion of the first paper, to our surprise, was that the amount of literature being produced in parasitology is very small. The total amount produced in those 19 major journals was the equivalent of one major journal in immunology. That is a little frightening.

When we looked at the subject matter in many of the classical journals in parasitology and tropical medicine, we found that the disciplines of the new biology were not being included.

Goffman's study of the citation analysis of the parasitology literature suggests it is much more isolated from the new biology. Parasitology is not quoted in the new biology journals, even at the 1% level, and the parasitology journals don't quote the new biology as frequently as they do virology and bacteriology. Comparing parasitology to virology and bacteriology, using this technique, its literature is much more isolated from the advances of new biology.

Another point is Paul Weinstein's that there is no parasitology textbook that is challenging enough conceptually to bright young people who want to go into the field. I believe we are at a point where work on such a textbook might begin; it is very important that people be encouraged to do something about that. I would like to see one as good as James Watson's *Molecular Biology of the Gene.* But that is so excellent it would be hard to beat; it is the best textbook I have ever read.

STUART: You are currently writing a book, aren't you?

WARREN: We are doing a new textbook of tropical medicine, not of parasitology. The big problem in doing a textbook of tropical medicine is to recruit people who have had a lot of experience in dealing with medicine in the tropics, as well as those who are relatively close to the implementation of modern science.

As an example, for the chapter on Chagas' disease we have Coura, a professor of tropical medicine in Rio de Janeiro, who is collaborating with Nadia Nogueira at Rockefeller University. For the chapter on African sleeping sickness, Tony Duggan is doing the clinical and Stephen Hajduli and Paul Englund are writing the basic part. That is an experiment, and we are waiting to see how it comes out.

CROSS: Are the existing publications adequate for purposes of disseminating information? Is there a need for new journals?

WARREN: More articles have been published in the parasitology journals than in the major immunology journals, but the delays in publishing in the *Journal of Immunology* are much shorter than in the parasitology journals. I think that anyone wishing to publish a paper on immunology of parasitic infections would prefer to see it in the *Journal of Immunology* than in the *American Journal of Parasitology.*

One interesting thing we noticed was that in 1979, 23 papers on schistosomiasis mansoni appeared in the *Journal of Immunology.* The most recent issue of that journal had three papers on schistosomiasis and one each on filariasis, malaria, and leishmaniasis. The top immunology journals are now publishing more papers on parasitology than ever before. Cross and Capron have frequently published in those journals, so the opportunities do exist.

I believe the parasitology journals have a useful function, and they have turned out to be quite reasonable publications, but I would rather publish in the *Journal of Immunology.*

OGLIVIE: That is a bit biased because, as far as British parasitology is concerned, we have been publishing in journals such as *Immunology* and *Clinical and Experimental Immunology* since the mid-1960s. I have analyzed the papers published in these journals in the past ten years and this has shown that there has been quite a high percentage of papers on parasitology in these journals for some time.

WARREN: I didn't mention them, but you are right. While I did mention the *Journal of Immunology,* other immunology journals are also publishing papers on parasitology, and that is a very real advance.

MITCHELL: One of the dilemmas is getting the cutting edge in a textbook on parasitology. There might be some difficulty in incorporating it into a textbook because, by definition, the cutting edge is very flexible and is likely to change. This year's breakthrough is next year's artifact; and one might not put it in a textbook. The cutting edge will never be found in a textbook because of publication delays and the fluctuating nature of publishing.

WARREN: I stand corrected. You are right; I shouldn't have said the "cutting edge," but I certainly think it is correct to say, "modern biology."

MITCHELL: I would agree with that.

DAVID: A lot of molecular biology came out of microbiology and virology; so obviously there is a cross-citation for virology and molecular biology that one wouldn't expect.

WARREN: That is true.

OGLIVIE: Why do you want to put parasitology into a textbook?

WARREN: Textbooks are a major form of communication for educators. That is something I would like to to do away with, but as long as we do teach that way, it is probably better to teach from the primary literature.

CROSS: Comparing textbooks on parasitology and microbiology, the field moves fast and the textbooks can become irrelevant.

SOPRUNOV: Our problem is that we cannot publish because we lack the money to do it. The Southeast Asian Information Center in Japan is trying to help the developing countries to publish textbooks and manuscrips.

CAPRON: Another serious problem for French-speaking scientists is the almost exclusive use of English as the universal scientific language. A good number of papers are published in French, which are never quoted in English-language journals.

I do not experience that any more because, like many French scientists, I learned the lesson and most of our papers are now published in English. But we might not be able to do so in the future because our government has set as a priority the defense of French language, and will link financial support to the publication of a certain number of papers in French.

WARREN: There is a problem of universal scientific language, and it does not involve only French. We have to think in terms of Spanish, Portuguese, and several other languages. The tendency to base the criteria of quality only on English-language journals and books is extremely dangerous.

But it would have been impossible for us to attempt to do an analysis of textbooks in every language in the world. In general, it is quite clear that if one uses the English language as an example, because it has tended to be the universal language of science in this part of the 20th century, the quality of the output in the journals and in the textbooks is higher.

BOOTH: I strongly support the French viewpoint because I have had the experience of editing a journal over a six-year period and of dealing with unhappy French writers because English and American writers do not refer to their work. Regardless of what people on the other side of the Atlantic think of the European literature, it is a real problem for the French. The Soviet Union has the same problem. Much good work is published there that is not referred to because people don't understand Russian. The Scandinavians publish in English, but that is an offense to the dignity of the individuals who speak Scandinavian languages. How do we solve that problem?

Warren: That would be a subject for another symposium.

16

Toward the Development of New Drugs for Parasitic Diseases

Fred R. Opperdoes

The International Institute of Cellular and Molecular Pathology (ICP), which was founded in 1974 in Brussels by Christian de Duve, has as its specific aim to promote close collaboration between the basic and applied medical research fields and their application to medicine and therapeutics through the great achievements, technologies, and knowledge gained during the last 25 years in the field of basic cellular and molecular biology. Being an officially recognized WHO Collaborating Centre, the ICP has set itself two tasks: (1) to contribute to the development of entirely new drugs by increasing our still poor knowledge about the causative agents of tropical diseases and (2) to improve the efficacy of already existing drugs by developing vehicles capable of targeting these to the tissues and organs affected by the parasite.

Trypanosomatid Hemoflagellates

Trypanosomatids are the causative agents of a number of serious diseases of man and his domestic animals. Human trypanosomiases, sleeping sickness (*Trypanosoma rhodesiense* and *T. gambiense*) in tropical and subtropical Africa, and Chagas' disease *(T. cruzi)* in tropical and subtropical areas of the Americas affect many millions of people. Trypanosomiases of livestock (Nagana) are caused by *T. brucei, T. congolense,* and *T. vivax* in tropical Africa and are responsible for the death of 3 million heads of cattle each year.

Human leishmaniasis is caused by four trypanosomatid species: *Leishmania tropica, L. mexicana, L. braziliensis,* and *L. donovani.* It is estimated that in total a hundred million people suffer from leishmaniasis.

In general, if untreated, most of the diseases caused by the trypanosomatids run a fatal course. Drugs when available are not very effective. They are too toxic or their use is restricted because of the spreading occurrence of resistance of the parasite against the drugs. Trypanosomiases are clearly a class of disease in which new and better drugs are needed; however, in the last 20 to 30 years no new drugs have been developed for the treatment of African trypanosomiasis. For Chagas' disease, so far, there exists no effective treatment at all.

We have chosen *Trypanosoma brucei* as a model organism for the study of the biochemistry and the cell biology of this group of parasites.

Glycolysis

In the bloodstream form of the African trypanosome *(T. brucei),* the single mitochondrion has been reduced to a peripheral canal containing no cytochromes and no functional Krebs' cycle.[1] Therefore, the bloodstream form is entirely dependent on glycolysis for its production of energy and in the bloodstream glucose is the preferred energy source. The trypanosome relies entirely on an exogenous source of carbohydrate. Fructose, mannose, and glycerol can replace glucose as energy source and they all support motility and respiration. Removal of exogenous substrate results in a rapid loss of respiratory activity and motility and the cells disintegrate. This indicates that the bloodsteam form is not able to survive for long on polysaccharide reserves or "high energy phosphate" stores like creatine phosphate, if at all present.

Glucose metabolism in *Trypanosoma brucei* bloodstream forms differs from glucolysis in other eukaryotes in a number of respects: (1) lactate dehydrogenase is absent and, therefore, the reducing equivalents generated in glycolysis are indirectly reoxidized by molecular oxygen via a dihydroxyacetone phosphate glycerol-3-phosphate shuttle plus a terminal oxidase: glycerol-3-phosphate oxidase[2]; (2) pyruvate is the end product of glycolysis and is excreted into the hosts bloodstream; (3) under anaerobic conditions glucose is quantitatively converted into equimolar amounts of pyruvate and glycerol, which are excreted.[3] This glucose dismutation proceeds with net ATP synthesis,[4] an observation which could not readily be explained until recently (see below).

How vulnerable a trypanosome is to inhibition of the glycolytic pathway is shown by the combined effect of salicylhydroxamic acid and glycerol, inhibitors of the aerobic and anaerobic glycolytic pathways, respectively. A single administration of this drug combination *in vitro* blocks glycolysis completely[5] and *in vivo* eliminates the trypanosomes from the bloodstream of the infected host within minutes.[6]

Glycolysis in *T. brucei* proceeds at an extremely high rate: 50% of its own weight in glucose is consumed per hour. The rate of glycolysis is directly coupled to the rate of ATP production[7] and recent calculations have indicated that such a high rate of glycolysis is required to enable the trypanosome to divide every seven hours.[8] Therefore, even an incomplete inhibition of glycolysis must lead to an increase in division time of the trypanosome and would help the host to overcome the infection by its natural immune system.

Another advantage of developing inhibitors of the glycolytic pathway of the bloodstream trypanosome might be the following: a drug that interferes directly with the energy production of the organism will kill the organism immediately. This drastically reduces the chance that the organism will develop resistance against such a drug.

The Glycosome

Recently it was discovered that in the bloodstream form of *T. brucei*, a number of glycolytic enzymes involved in the conversion of glucose and glycerol into 3-phosphoglycerate are localized in special membrane-bounded organelles, which we have called "glycosomes"[9] (Fig. 16-1). The presence of these glycosomes, which have a microbody-like appearance, is not a special form of adaptation of *T. brucei* to its stay in the glucose-rich bloodstream of its host, since glycosomes have also been found in cultured procyclics of *T. brucei*,[10] in *Crithidia* spp.[11] and in different stages of the life cycle of *T. (S) cruzi* and *Leishmania mexicana*.[12, 13] This makes the presence of glycosomes a general property of the trypanosomatids. Therefore, any drug that will be developed as a specific inhibitor of glycolysis in *T. brucei* bloodstream forms might also be effective against the causative agents of leishmaniasis and Chagas' disease.

The presence of a number of glycolytic enzymes inside an organelle, as has been found in trypanosomatids, is a property unique to these organisms. In any other organism studies so far, prokaryote or eukaryote, the glycolytic reactions proceed in the soluble portion of the cell. This, for trypanosomatids' unique property, makes it likely that the glycolytic enzymes, present inside the glycosome, exhibit physical and chemical characteristics which are different from their soluble counterparts present in the host cells, which would render them excellent targets for future chemotherapy. In fact, hexokinase and phosphofructokinase from *T. brucei* have kinetic and regulatory properties which are quite different from their mammalian counterparts.[14] Apart from the fact that in trypanosomes the glycolytic enzymes are located in a glycosome, we have recently obtained evidence that even inside the glycosome a number of the enzymes are associated with each other. In contrast to the interactions between glycolytic enzymes that have been described for the enzymes of bacteria and higher eukaryotes, which are weak, the interactions between

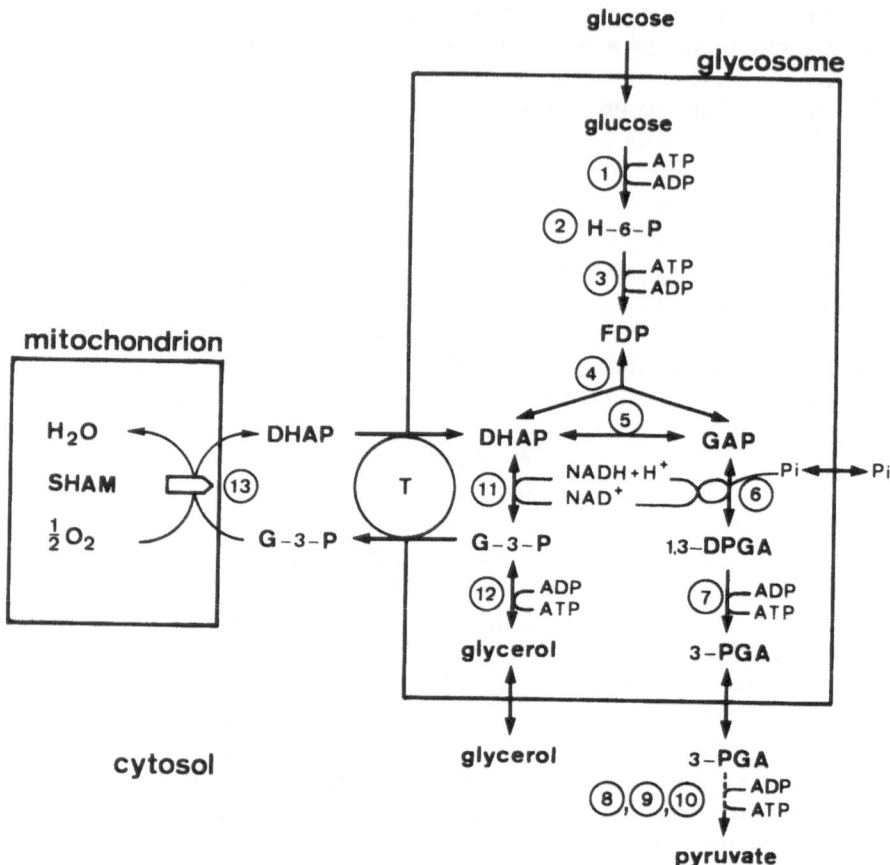

FIG. 16-1. Compartmentation of glycolysis in *Trypanosoma brucei:* 1, hexokinase; 2, glucose-P-isomerase; 3, p-fructokinase; 4, aldolase; 5, triose-P-isomerase; 6, glyceraldehyde-P-dehydrogenase; 7, 3-P-glycerate kinase; 8, P-glycerate mutase; 9, enolase; 10, pyruvate kinase; 11, glycerol-3-P dehydrogenase; 12, glycerol kinase; 13, glycerol-3-P oxidase T, DHAP: glycerol-3-P translocator.

the trypanosoma enzymes are strong. Irrespective of the homogenization procedure or of the detergent used, the majority of the enzymes behave as members of a big multi-enzyme complex.[15,16]

Anaerobic Glycolysis

Several hypothetical schemes have been proposed to account for the fact that trypanosomes are capable of ATP synthesis under anaerobic conditions. Additional glycolytic enzymes[8,9] were suggested but none of these was found. With the discovery of the glycosome, Opperdoes and Borst[9] hypothesized that the production of glycerol with the concomitant synthesis of ATP had to be the result of special conditions existing inside the glycosome under anerobiosis, and they proposed that these conditions led to a reversal of the glycerol kinase catalyzed reaction. Recently evidence was obtained that such conditions might exist inside the glycosome and that glycerol-3-phosphate is the direct precursor of glycerol.[17-19] Hammond and Bowman[20] described a glycerol-3-phosphate: ADP transphosphorylase activity which could be attributed to glycerol kinase, and they measured in pulse-labeling experiments with [14C]glycerol[18] that under anaerobiosis glycerol is incorporated into glycerol-3-phosphate at a much higher rate than the net glycerol production (Fig. 16-2). Since glycerol-3-phosphate is not metabolized further under anaerobiosis, this must mean that the glycerol kinase reaction is at equilibrium, thus confirming the hypothesis of Opperdoes and Borst.[9]

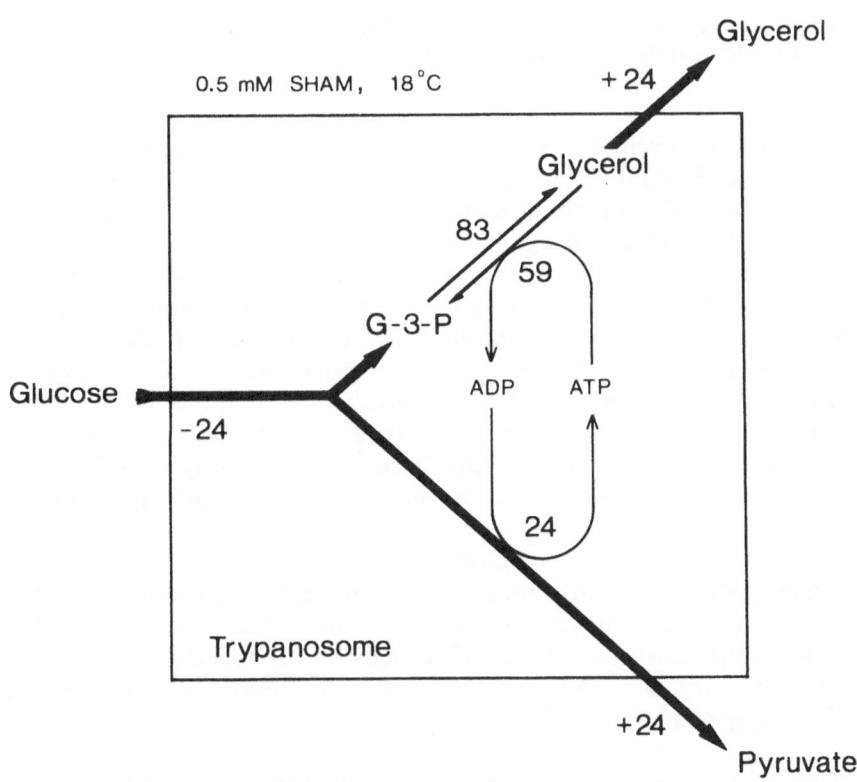

FIG. 16-2. The involvement of glycerolkinase in anaerobic glycolysis of *T. brucei.* Drawn from data by Hammon and Bowman. Figures indicate the rate of metabolite fluxes in nmol min^{-1} mg^{-1} protein.

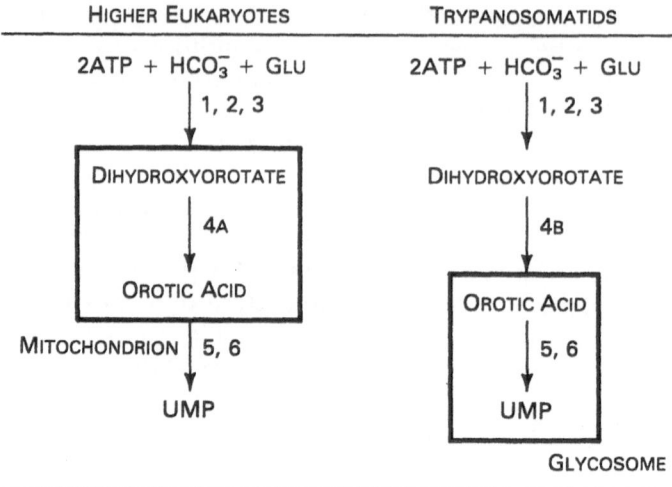

FIG. 16-3. Comparison of the compartmentation of pyrimidine biosynthesis in try-panosomatids with that of other eukaryotes.

Other Pathways Associated with the Glycosome

The first three enzymes of the pathway of pyrimidine biosynthesis are present in the cytosol (Fig. 16-3), as is the situation in mammalian cells. The fourth enzyme, being a dihydro-orotate dehydrogenase in host cells and associated with the mitochondrion, is a dihydro-orotate oxidase in trypanosomatids and is also present in the cytosol. In all organisms studied to date, the last two enzymes of the pathway, orotate phosphoribosytransferase and orotidine 5'-phosphate decarboxylase, are soluble enzymes except in trypanosomatids, where they are firmly membrane-bound. We have recently been able to show that both these enzymes are associated with the glycosomes of *T. brucei*,[21] and there are now strong indications that the same situation exists in *T. cruzi* and in the leishmanial parasites.

Conclusion

The data presented clearly indicate that the trypanosomatids have some unique properties resulting from the presence of the glycosome. Therefore, it is imperative that these differences be exploited for the development of new trypanocidal drugs. Features that lend themselves well for further study are:

1. The mechanism of glucose transport into the trypanosome.
2. The properties of the individual glycolytic enzymes and the development of specific inhibitors.
3. The development of an inhibitor of glycerol kinase in combination with an inhibitor of respiration.

4. The properties of the glycosomal membrane and the development of compounds that might interfere with its integrity.
5. The properties of the hypothetical carrier of glycolytic intermediates in the glycosomal membrane.
6. The properties of the last three enzymes of the pathway of pyrimidine biosynthesis.

At present points 1, 2, 4, and 5 are pursued in our laboratory with as ultimate goal the development of a more selective chemotherapy.

Drug Targeting

One of the projects carried out in the ICP under the guidance of André Trouet is the chemotherapy of cancer. The main interest of his group is the development of vehicles that can be used for the targeting of drugs toward malignant cells. Due to the very nature of a number of infections caused by tropical parasites, however, their approach might also be applicable to various tropical diseases.

In many infections caused by protozoan parasites, the causal agent survives and proliferates intracellularly, most often within the vacuolar system into which it entered via a mechanism of phagocytosis. The infected host cells are mainly the polymorphonuclear leukocytes and macrophages, but other cell types like hepatocytes and striated muscle may also be involved depending on the type of parasite and the host. After recognition has taken place between parasite and the outer surface of the host cell, most likely mediated by receptors present on one or both cells involved, the plasma membrane of the host cell invaginates around its prey and traps it inside a vacuole called a "phagosome" (Fig. 16-4). The normal fate of such a phagosome is to fuse with the lysosomes of the cell by which the hydrolytic enzymes contained within these vacuoles come into contact with the prey and digest it. If this occurs the infectious agent is killed. The host cell may, however, fail to kill the parasite for several reasons, because (1) the parasite escapes from the phagosome or parasitophorous vacuole into the cytoplasm before fusion with the lysosomes occurs, as in the case with *Trypanosoma cruzi;* (2) the parasite succeeds in preventing lysosomal phagosomal fusion, as is the case with *Toxoplasma* spp., or (3) the parasite is capable of surviving within the hostile environment of lysosomes, where it resists or inactivates the hydrolytic enzymes, as in the case with the amastigotes of *Leishmania* spp. The result is that the parasite survives and starts to multiply. The host cell then bursts open and liberates numerous parasites which penetrate into neighboring cells and so the infection is spread.

Most of the drugs used in the treatment of protozoal infections are rather toxic and active only in a dose close to the one which is maximally toler-

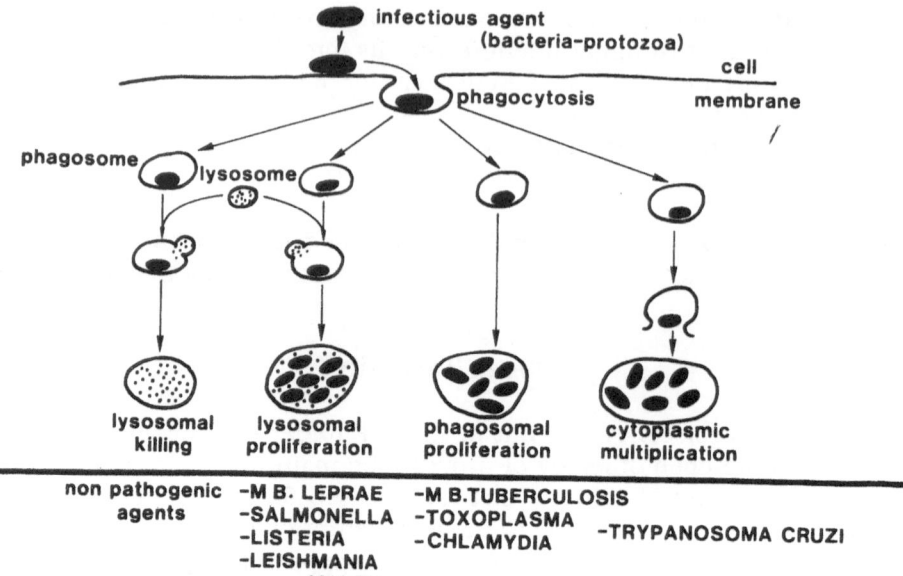

FIG. 16-4. Intracellular host–parasite relationships.

ated. If, however, a drug could be directed toward the infected cells, or tissues, without diffusing into highly sensitive but otherwise healthy organs, the efficacy of such a drug would be significantly increased, whereas the toxicity would be decreased and consequently a much better result would be obtained at a lower dose.

Possible carriers for targeting such drugs could be proteins, which when linked to drugs can be recognized by specific receptors present on the outer surface of the target cell. Such proteins are then interiorized and subsequently concentrated in the lysosomes where digestion follows and the drug is released. This kind of approach, which makes use of the same principles by which the parasite entered the host cell, requires that the drug-carrier complex have a number of important properties, which are summarized in Table 16-1.

Chemotherapy of Rodent Malaria

I would like to show with an example how the principle of drug targeting was applied to malaria caused by *Plasmodium berghei* in mice. In rodent malaria, as is the case in human malaria, the sporozoites, after their inoculation into the bloodstream by an infected mosquito (Fig. 16-5), invade the hepatocytes of the liver and subsequently develop into "schizonts." These schizonts release "merozoites," which then infect the red blood cells. Some human plasmodial parasites, especially *P. vivax* and *P. ovale,* are capable of surviving in the liver for long times. These inactive dormant

TABLE 16-1. Criteria to be fulfilled by a lysosomotropic drug-carrier conjugate

---

I. The carrier should be
   a. selective for the target cell,
   b. endocytized by the target cell and transferred into the lysosomal compartment,
   c. biodegradable,
   d. nonimmunogenic,
   e. capable of permeating through anatomical barriers between administration site and target.
II. The drug should resist lysosomal pH and lysosomal enzymes.
III. The drug-carrier conjugate should be
   a. stable in bloodstream and extracellular fluids,
   b. pharmacologically inactive,
   c. sensitive to lysosomal enzymes or acidic pH and release the drug in an active form,
   d. nonimmunogenic.

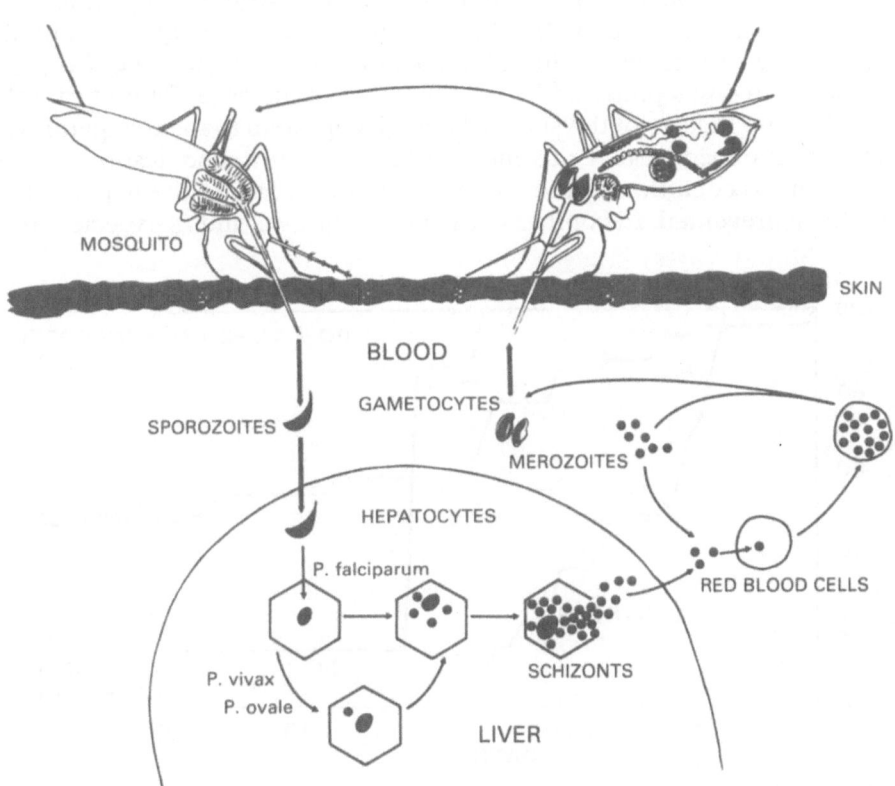

FIG. 16-5. The life cycle of malaria parasites.

stages, which are called "hypnozoites," can give rise to a recurrence of the disease even after all parasites have been cleared from the blood. A drug active against these liver stages of the parasite is primaquine which, in combination with other drugs directed against the blood stages, is used to obtain complete cures. The toxicity of primaquine, however, greatly restricts its usefulness as a causal prophytactic.

In our laboratory Philippe Pirson together with Michèle Masquelier and Roger Baurain have linked primaquine (PQ) to asialofetuin, a glycoprotein with terminal galactose residues, which is selectively recognized and taken up by the hepatocytes of the liver. As linker between drug and protein, they used a tetrapeptidic spacer arm which consisted of the amino acids Ala-Leu-Ala-Leu (AA$_4$), which has proven to be stable in the bloodstream but digestible by lysosomal hydrolases of liver, thus releasing free PQ from the conjugate as soon as it has reached the lysosomes.[22,23] According to this principle two conjugates were prepared. One—PQ-(AA$_4$-S)-glycoprotein—involving a minimal modification of the protein part, was synthesized by succinylation of the terminal amino group of the tetrapeptidyl-PQ prior to its linkage to the asialofetuin. This conjugate, when administered IV three hours after infection of the mice with *P. berghei* sporozoites, allowed the cure of all animals with a single dose equivalent to 25 mg free PQ diphosphate (Fig. 16-6). The other conjugate—PQ-(AA$_4$)-S-glycoprotein—was prepared by succinylation of the asialofetuin first, followed by the linkage of the succinylated glycoprotein to the tetrapeptidyl-PQ. It was expected that this method led to such a modification of the carrier that recognition by the receptors on the surface of the hepatocytes would be prevented. In contrast to the first conjugate, and as expected, the

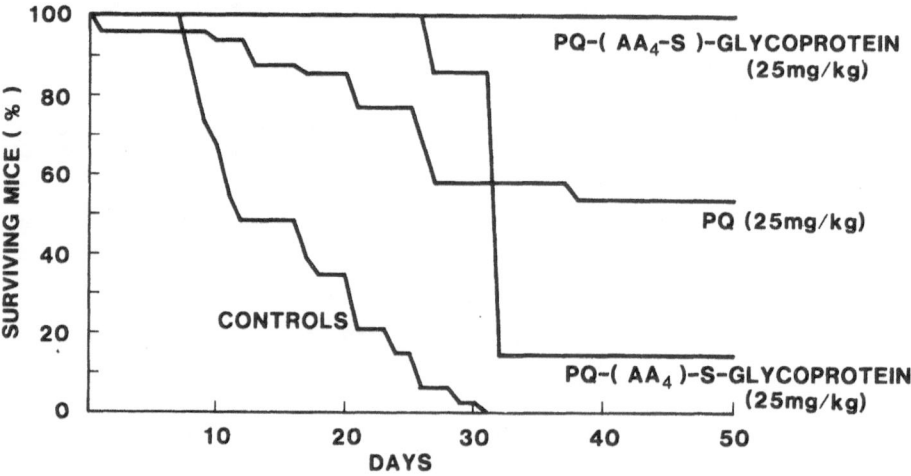

FIG. 16-6. *In vivo* activity of primaquine asialofetuin conjugates against rodent malaria.

second conjugate was much less active and even less active than the free PQ. The latter result suggests strongly that when linked to a carrier which has lost its specificity for the hepatocytes, PQ is taken up less by these cells than when given as the free drug. The dramatic difference in cure obtained with the two conjugates indicates that recognition of the glycoprotein part of the conjugate by hepatic receptors plays a vital role in the targeting of PQ.

Conclusion

The results described provide us with the first evidence in favor of the possibility of targeting drugs under *in vivo* conditions. This kind of approach, however, should not be restricted to hepatocytes alone. In principle it can be extended to many other cell types provided that they are capable of endocytosis and contain receptors specific for proteins that have to be taken up by these cells. Numerous infections caused by protozoa, viruses, or bacteria which invade cells like hepatocytes or those of the reticulo-endothelial system would lend themselves well for a treatment with similar drug-carrier conjugates.

References

1. Bowman IBR, Flynn IW (1976) Oxidative metabolism of trypanosomes. In: Lumsden WHR, Evans DA (eds) Biology of the kinetoplastida. Academic Press, New York, Vol I, pp 435–476
2. Grant PT, and Sargent JR (1960) Properties of L-α-glycero-phosphate oxidase and its role in the respiration of *Trypanosoma rhodesiense*. Biochem J 76: 229–237
3. Ryley JF (1956) Studies on the metabolism of protozoa. 7. Comparative carbohydrate metabolism of eleven species of trypanosomes. Biochem J 62: 215–222
4. Opperdoes FP, Borst P, Fonck C (1976) The potential use of inhibitors of glycerol-3-phosphate oxidase for chemotherapy of African trypanosomiasis. FEBS Lett 62: 169–172
5. Fairlamb AH, Opperdoes FR, Borst P (1977) New approach to screening drugs for activity against African trypanosomes. Nature (London) 265: 270–271
6. Clarkson AB, Brohn FH (1976) Trypanosomiasis: An approach to chemotherapy by the inhibition of carbohydrate catabolism. Science 194: 204–206
7. Fairlamb AH, Bowman IBR (1980) Uptake of the trypanocidal drug Suramin by bloodstream forms of *Trypanosoma brucei* and its effect on respiration and growth *in vitro*. Mol Biochem Parasitol 1: 315–333
8. Brohn FH, Clarkson AB Jr (1980) Pattern of glycolysis at 37°C *in vitro*. Mol Biochem Parasitol 1: 291–305
9. Opperdoes FR, Borst P (1977) Localization of nine glycolytic enzymes in a microbody-like organelle in *Trypanosoma brucei* the glycosome. FEBS Lett 80: 360–364

10. Opperdoes FR, Markoš A, Steiger RF (1981) Localization of malate dehydrogenase, adenylatekinase and glycolytic enzymes in glycosomes and the threonine pathway in the mitochondrion of cultured procyclic trypomastigotes of *Trypanosoma brucei*. Mol Biochem Parasitol 4: 291–309
11. Opperdoes FR (1981) Subcellular compartmentation. In: Klein RA, Miller PGG (eds). Alternate metabolic pathways in protozoan energy metabolism. Parasitology 82: 1–30
12. Taylor MB, Berghausen H, Heyworth P, Messenger N, Rees LJ, Gutteridge WE (1980) Subcellular localization of some glycolytic enzymes in parasitic flagellated protozoa. Int J Biochem 11: 117–120
13. Coombs GM, Craft JA, Hart DT (1982) A comparative study of *Leishmania mexicana* amastigotes and promastigotes. Enzyme activities and subcellular locations. Mol Biochem Parasitol 5: 199–211
14. Nwagwu M, Opperdoes FR (1982) Regulation of glycoloysis in *Trypanosoma brucei*: hexokinase and phosphofructokinase activity. Acta Trop 39: 61–72
15. Opperdoes FR, Nwagwu M (1980) Subcellular localization of glycolytic enzymes in the glycosome of *Trypanosoma Brucei*. In: Van den Bossche H, (ed) The host-invader interplay. Elsevier/North Holland Biomedical Press, Amsterdam, pp 683–686
16. Oduro KK, Bowman IBR, Flynn IW, (1980) *Trypanosoma brucei*: preparation and some properties of a multi-enzyme complex catalysing part of the glycolytic pathway. Exp Parasitol 50: 240–250
17. Visser N, Opperdoes FR (1980) Glycolysis in *Trypanosoma brucei*. Eur J Biochem 103: 623–632
18. Hammond DJ, Bowman IBR (1980) *Trypanosoma brucei*: the effect of glycerol on the anaerobic metabolism of glucose. Mol Biochem Parasitol 2: 63–75
19. Visser N, Opperdoes FR, Borst P (1981) Subcellular compartmentation of glycolytic intermediates in *Trypanosom brucei*. Eur J Biochem 118: 521–526
20. Hammond DJ, Bowman IBR (1980) Studies on glycerol kinase and its role in ATP synthesis in *Trypanosoma brucei*. Mol Biochem Parasitol 2: 77–91
21. Hammond DJ, Gutteridge WE, Opperdoes FR (1981) A novel location for two enzymes of de novo pyrimidine biosynthesis in Trypanosomes and *Leishmania*. FEBS Lett 128: 27–29
22. Trouet A, Masquelier M, Baurain P, Deprez-De Campeneere D (1982) A covalent linkage between daunorubicin and proteins that si stable in serum and reversible by lysosomal hydrolases, as required for a lysosomotropic drug-carrier conjugate: *in vitro* and *in vivo* studies. Proc Natl Acad Sci 79: 626–629
23. Trouet A, Pirson P, Steiger R, Masquelier M, Baurain R, Gillet J (1981) Development of new derivatives of primaquine by association with lysosomotropic carriers. Bull WHO 59: 449–458

17

The Role of Molecular Biology in Parasitology

George A. M. Cross

Molecular biology is perceived by most people as having its effective begining in 1953, with the famous publication by J. D. Watson and F. H. C. Crick of a proposal for the structure of DNA.[1] In truth, the field had its real beginings a little earlier. Yet, by any measure, molecular biology is a young discipline—young enough indeed for the persistence of abundant discussion and controversy concerning its origin, definition, and the personalities that dominate it. Some would assert that it has no right to be regarded as a separate discipline. Molecular biology, as the term is generally understood, had its origin in microbial genetics and is perhaps best seen as an attempt to fuse the experimental and intellectual approaches of genetics and biochemistry without being conscious of doing so. This was, of course, regarded by some, especially the traditional biochemists, as something akin to a circus act.

The history of molecular biology has been most elegantly, thoroughly, and entertainingly documented by Judson in his book *The Eighth Day of Creation.*[2] I shall only mention here that the controversy of the definition of molecular biology and its practitioners continues, as can be seen from recent articles by Sibatani[3] and Kurland.[4] Much of the defensive/aggressive reaction of classical biological scientists can be attributed to fears of incipient megalomania amongst molecular biologists, fears fueled by remarks such as "Molecular biology can be defined as anything that interests molecular biologists," which Judson attributes to Crick.

For present purposes I propose that we adopt a more functional definition, namely, the study of biological phenomena at the molecular level. Perhaps from this standpoint we can best understand how the more recent developments in molecular biology can offer the most relevant opportunities to parasitology. Although parasites are fascinating organisms in themselves as well as presenting, in some cases, intriguing models of normal or deviant genetic and metabolic behaviour, the major concern of this paper is the opportunities which molecular biology may present for the alleviation of the current global dominance of parasitic diseases, in the sense of those organisms that are the domain of the traditional field of parasitology.

In this more narrow area of practical opportunities, the most relevant aspects of molecular biology are the so-called new technologies of gene cloning and monoclonal antibodies. The essential tools comprising these technologies arose in the course of fundamental biological research aimed solely at understanding basic molecular and genetic mechanisms.

Monoclonal Antibodies

How new are the monoclonal antibody and recombinant DNA technologies? The crucial experiment that generated the first monoclonal antibodies[5] was an extension of studies on the fusion of mutant myeloma cell lines which were initiated with a view to extending the current understanding of the regulation of immunoglobulin gene expression. The potential uses that were quickly recognised for monoclonal antibodies are limitless. They have already found widespread use in defining cell populations and for exploring cellular structures by immunocytochemical techniques. In the near future, we expect to see monoclonal antibodies being used in a wide range of serological tests, where they have advantages over conventional antisera. In the field of parasitology, as elsewhere, they will have important applications in taxonomy and epidemiology, and they may, in the form of anti-idiotype monoclonal antibodies, act as surrogate antigens in either immunodiagnostic tests or even in vaccines.

The advantages of monoclonal antibodies derive from three basic properties of the technique: first, monoclonal antibodies can have exquisite specificity; second, that specificity can be obtained without the need for extensive prior purification of an antigen; third, a monoclonal antibody is a defined chemical with constant properties. Monoclonal antibodies of different classes and subclasses and of differing affinities and with specificities for different epitopes on the same antigen may be produced. In the event that the specificity of a monoclonal antibody itself might be too narrow, the monoclonal can be advantageously employed in the purification of an antigen which can then be used to produce a highly specific polyclonal antiserum. Logically, though perhaps some would think it perverse, monoclonal antibodies can sometimes appear which show much wider cross-reactions than normal antisera. This may happen because the selected monoclonal antibody is one that would normally be very poorly represented in a conventional polyclonal antiserum. So, one must be alert to that possibility and its significance or otherwise when interpreting results.

Production Techniques

In principle, the techniques for producing monoclonal antibodies are simple and have been well described in the literature. They are labor-intensive and require meticulous cell culture technique. The primary drawback derives from the random nature of the essential events involved in the practical realisation of the conceived qualities of the antibody which is being sought in each situation. As with most biological techniques, there is no substitute for experience, and the best course of action is to take advice from successful exponents of the techniques. If you are starting from scratch, be sure to get your myeloma cells from a reliable, authenticated source. They must be growing well to fuse well. Given the relative inefficiency in generating hybrids (about $1/10^5$ spleen cells), it is important to maximise antibody production by the donor spleens. This may be achieved by immunization in one of several ways, ranging from injection of purified or enriched antigens, disrupted cell homogenates, fixed or irradiated cells, or live infections which can generate natural immunity. Most fusions have been performed after one or a few immunizations using different protocols, with or without adjuvants as appropriate, but almost all involve a final intravenous injection of antigen three to five days prior to fusion. Antibody titer should be determined prior to fusion and spleens from the highest titer mice should be selected for fusion.

In most successful fusions, several hundred antibody-positive culture wells will be found in the initial screening of hybrids. As the labor involved intensifies as cultures are expanded, cloned, and screened, only 10–20 positive wells can generally be selected from the initial screening. Unfortunately, on average, 50% of the initially selected hybrids will be lost due to instability, poor growth, or cessation of secretion during subsequent handling up to and including growth as ascitic tumors. It is therefore most important to use appropriate screening techniques at this early stage of hybridoma selection. The method chosen to screen fusions should reflect the properties required of the desired hybrid. This may seem obvious, but it is frequently forgotten in the desire for convenience in screening. It is especially important in the selection of monoclonals, which will ultimately be required to form the basis of rapid, convenient, reliable, specific, and sensitive immunodiagnostic tests. Preferably two screens should be used, in parallel not sequentially, to optimize the selection of appropriate hybrids.

Use in Parasitology

The main relevance of this brief exultation on the subject of monoclonal antibodies is in their unsurpassed qualities as probes of cellular structure at the molecular level. The most advanced and exciting use of monoclonal antibodies in parasitology so far is in the identification of protective antigens from malaria parasites. Essentially similar approaches to antigen characterisation are being pursued using a variety of malaria parasites. A stage and species-specific 44,000 MW protein (Pb44) was recognisable by immunoprecipitation with conventional antiserum as being possibly the major antigen on the outer surface of *Plasmodium berghei* sporozoites. A monclonal antibody was, however, necessary to prove, by its exquisite activity in serum transfer experiments, the relevance of Pb44 to protection.[6,7] Holder and Freeman have been able to take their work with the blood stages of *P. yoelii* one stage further, by successfully immunizing mice with two antigens (235,000 and 230,000 MW) identified and subsequently purified using monoclonals.[8]

The only monoclonal work which has so far been published on *P. falciparum* blood stage antigens[9,10] has relied extensively on merozoite invasion inhibition to identify potentially relevant polypeptides of 96,000, 36,000 and 41,000 MW. A monoclonal antibody precipitating a 195,000 MW protein did not inhibit invasion. Monoclonal antibodies which block invasion of erythrocytes by *P. knowlesi* have also been reported.[11] These antibodies precipitated a 250,000 MW protein. It is not yet known whether the antigens identified on the various species of Plasmodium have any common features. Anti-gamete transmission-blocking monoclonal antibodies recognising antigens on the surface of macro- and microgametes of *P. gallinaceum* have also been identified.[12]

The power of monoclonal affinity chromatography for antigen purification is well illustrated by the *P. yoelii* work. Starting with detergent-solubilised whole blood, two passes over affinity columns produced purifications of 2,500-fold and 11,000-fold for the two antigens of interest (A. A. Holder, personal communication). The amount of material required for immunization studies currently makes these a practical proposition only with bloodstream stages and, even then, the effort and expense involved is high. It seems likely that such direct immunization tests will not be done on genuine parasite material in the case of sporozoite and gamete antigens, but will be achieved through gene cloning or synthetic peptide synthesis.

Recombinant DNA

That seems a good cue to return to the subject of gene cloning or recombinant DNA technology. It is rather more difficult to identify a key starting point for the recombinant DNA technology which has become possible only through a series of major developments in techniques for the manipulation and characterisation of DNA. If any single event provided the key to gene cloning it was the isolation, in 1970, of the first sequence-specific DNA-cutting enzyme (restriction endonuclease). This led, in 1972–1973, to the first precise construction of recombinant DNA molecules *in vitro* and their insertion into bacteria. The years 1973–1976 were clouded by the great safety debate[13] but, thereafter, the pace of research, discovery, and publication in this field has grown exponentially. The first company to be founded exclusively for the commercial exploitation of recombinant DNA techniques, Genentech, was established in 1977 and a patent application for the bacterial synthesis of Human Insulin A and B chains was filed in the same year.

Safety considerations, with emphasis on biological containment of recombinant DNA molecules, led to the development of the bacteria, plasmid, and bacteriophage strains which are now in everyday use. Although *Escherichia coli* remains the workhorse of gene cloning, other bacterial, mammalian, and yeast host–vector systems are now available for the selection and expression of recombinant DNA. Other key developments have occurred in gene synthesis and in DNA sequencing. More than 300 restriction enzymes with over 85 useful distinct sequence specificities have now been characterised[14]; it is the widespread availability of these and other enzymes and techniques, together with the dissemination of the relevant expertise, which has made gene cloning the universal and powerful technique of today.

Gene cloning techniques give us the ability to select, purify, amplify, and characterise a gene, down to the level of the complete nucleotide sequence, and to produce the discrete product of that gene and to investigate the functions of that gene product (usually a protein) in either a normal, abnormal, or novel situation. Let us therefore first consider the ways in which cloned genes might be used in parasitology, as elsewhere, for practical purposes. I have tried to cover most of the likely uses in Table 17-1. The major application of gene cloning is in the elucidation of the genetic basis of normal and abnormal cell function. This is the objective of the true molecular biologist, from which all practical possibilities arise. The potential for using cloned DNA probes for the diagnosis of genetic disorders and the detection of disease-causing organisms has already found application in the prenatal detection of inherited abnormalities[15] and identification of microbial pathogens.[16] The production of several biologically active proteins, including insulin, growth hormone, and inter-

TABLE 17-1. Uses and advantages of cloned genes in medical and veterinary science

1. Elucidation of genetic basis of normal and abnormal cell function
2. Diagnosis of disease and disorders
3. Synthesis of biologically active proteins
4. Subunit vaccines may provide
   a. improved vaccine production
   b. improved vaccine efficacy
   c. new vaccine production
   d. side effects absent due to purity
   e. safety in manufacture and use
   f. consistency in manufacture and use

feron, from recombinant gene constructions expressed in bacterial and yeast cells has already received widespread attention, although it must be emphasised that several nontrivial problems have still to be overcome before these achievements can be fully implemented in medical practice.

Vaccine Production

Perhaps the greatest hope of parasitologists, and many others, is in the potential for the production of vaccines agaisnt diseases for which vaccines are not presently available and indeed are probably beyond realisation by the application of conventional techniques. Through gene cloning it may become possible to produce large amounts of an exact replica of a natural parasite antigen, as the essential vaccine ingredient. In some cases this may not be the necessary, desirable, possible, or most effective solution. For example, the large proteins already identified as putative vaccine entities in rodent and human malaria must have many antigenic sites, not all of which would be expected to be essential for the generation of a protective immune response. Even small synthetic peptides, synthesised through the knowledge and intelligent interpretation of natural protein sequences, now determinable most efficiently from a DNA sequence, may be effective immunogens when coupled to appropriate carriers, as has been extensively demonstrated by Ruth Arnon (see elsewhere in this volume) and her colleagues and more recently promoted by Lerner.[17] The production of either a synthetic peptide or a recombinant-derived protein, however, is only one step in the formulation of an effective vaccine. Both routes generally give a basic vaccine ingredient which is a much weaker immunogen than when presented in its natural state, as part of the infectious agent. We have therefore the important task of learning how to present our defined antigens efficiently, without carrier-mediated side effects, to the human patient. On the other hand, and here perversity may work to our advantage, an isolated antigen may be more effective in cases where the presence of other components in the natural agent leads to suppression of, or the generation of other impediments to, an efficient immune response in the natural infection.

Knowledge of the genetic structure and pathogenic mechanisms of infectious agents, through recombinant DNA studies, may also lead appropriately in some cases to the production of live vaccines based on rational attenuation. This may be particularly effective for the stimulation of immunity to intestinal agents. Thus, many laboratories are already working towards the production of safe vaccines against bacterial toxins, including cholera toxin, through a generalised approach of deleting genes or parts of genes responsible for toxicity whilst retaining the essential immunogenic determinants. This approach may also be applied to viruses and may be particularly appropriate where, as is apparently the case wtih the capsid proteins of poliovirus, it is currently impossible to obtain neutralising responses by vaccination with purified viral proteins. In the case of poliovirus, Racaniello and Baltimore[18] have demonstrated the infectivity of cloned DNA derived from the RNA genome, which shows the way to the construction of specifically modified RNA viruses. Techniques for the direct modification of RNA sequences are essentially nonexistent. These authors share my optimism that such approaches will, in some cases, play an important role in the development of new, safe, and effective vaccines.

Gene Cloning

This is not the place to detail the practical techniques of gene cloning. That subject has been well described elsewhere in several essays directed to the educated layman or non-molecular biologist.[13,19] There are several alternative strategies for gene cloning and many tactical variations. I do think, however, that it is appropriate here to summarize the main steps which would be involved in a gene cloning exercise, aimed at the ultimate target of efficiently expressing a putative vaccine antigen, in order to impart an understanding of some of the time factors, obstacles, and other key factors involved. In Table 17-2, I have made a breakdown of the main steps involved.

The requirements for gene cloning, apart from expertise, motive, and the identity of the protein whose gene is to be cloned, are as follows. First, it is desirable to have an enriched source of the gene. As can be seen from the information presented in Table 17-3, the genetic complexity of different biological organisms varies over a million-fold range. Thus, while it is a rather straightforward matter to clone and identify a desired gene from a

---

TABLE 17-2. Requirements for gene cloning

1. Motive, expertise, gene identifier
2. Source of desired gene (preferable enriched)
3. Strategy for isolation of desired gene
   a. enrichment,
   b. making copy DNA if necessary,
   c. splicing enzymes to insert DNA into
   d. vector: plasmid or phage capable of
   e. transformation of and replication in a suitable
   f. host (preferably a bacterium),
   g. selection,
   h. mapping, and
   i. sequence determination.
4. Determine expression objectives (fusion proteins, secretion or not, processing and other post-translational modifications)
5. Design and construction of an expression system to optimize expression (transcription-promotors etc., and translation)
6. Large-scale culture of recombinant cells, purification, and formulation of product

---

TABLE 17-3. Scale of genetic complexity in living organisms

	Small animal or bacterial virus or plasmid	Bacterium	Animal cell
Size of cell		1–2 μm	20–40 μm
DNA: MW	2.5×10^6	2.5×10^9	2.5×10^{12}
Base Pairs	4×10^3	4×10^6	4×10^9
Length	3 μm	3 mm	3 m
Proteins	6	3,000	60,000?
In a boardroom-size cell, DNA length would be:	30 m	20 miles	2,000 miles*

*Thickness would be 1 mm.

small animal or bacterial virus, the task becomes rather more difficult in a bacterial cell and extremely time consuming for a mammalian gene. Cloning bacterial genes using bacterial host–vector systems is relatively straightforward for several reasons, mainly attributable to the mutual compatibility of gene organisation, transcription, and translation control signals. This is not the case when cloning eukaryotic genes in bacterial host–vector systems. Apart from the thousand-fold greater amount of DNA in the nucleus of a mammalian cell, there are additional complications such as the widespread occurrence of split genes. The availability of an enriched source of the gene is therefore essential and, as many mammalian cells are differentiated to express only a small proportion of the available genes at any time, the enrichment we seek is usually found in the form of the messenger RNA (mRNA) in cells which are differentiated to produce meaningful amounts of the gene product. In some cases, tumors can provide an even greater enrichment of a specific mRNA. Even so, it will generally be desirable to do one stage of specific *in vitro* mRNA enrichment prior to cloning [as complementary/copy DNA (cDNA)] or use as a hybridisation probe of a "shotgun" clone bank. In eukaryotic systems, mRNA may also be the only form in which the genetic information is colinear with the protein sequence. As the total amount of cellular DNA in Plasmodium and trypanosomes is only three to six times greater than in *E. coli*,[20,21] the genetic organisation of parasites might be closer to bacteria than to mammalian cells, on the scale of complexity. If protein amino acid sequence information is available, the synthesis of a small gene or a small DNA sequence coding for an immunogenic peptide is quite a realistic undertaking today.

Following extraction and purification of mRNA, where necessary, the most difficult and time-consuming step in gene cloning is the identification of the required clone. Enrichment of the gene bank for the sequence of interest reduces the effort required for its identification, and so much time is often spent on prior RNA purification. A key tool for assessment of RNA purification is the cell-free protein synthesising system, usually derived from rabbit reticulocyte; our monoclonal antibodies or monospecific polyvalent antisera produced with monoclonal-purified antigens are useful here for the identification of specific RNA translation products. Unfortunately, antisera are unlikely to be useful as a direct primary screen for identifying the required bacterial clone. This is due to the many factors which operate against the likelihood of our gene being detectably expressed by the bacteria in which we are likely to do our initial cloning. The number of likely clones is narrowed down by using hybridisation probes of as high a purity as obtainable. The smaller panel of positives will be further screened by more definitive but more difficult and time-consuming methods such as hybrid-selected mRNA translation or oocyte assays. In some cases, differential hybridisation can provide an extremely effective short-cut to identification of desired recombinants, as was illustrated by the cloning of trypanosome VSG genes.[22] Synthetic oligonucleotides may also be used as hybridisation probes, if a suitable short amino acid sequence with minimal codon redundancy is known.

Once one convincing positive clone has been identified, it may, for example if it is incomplete, be used to identify further clones. The speed of progress accelerates following the definitive identification of the required clone. Information on related genes and chromosome organisation can be rapidly obtained. Extensive characterisation of the gene by restriction mapping and DNA sequencing is required for the gene to be manipulated into appropriate host–vector constructions that will permit expression of high levels of the corresponding protein. Again there is no assurance that a system which is known to express another nonbacterial protein in high amounts will do the same with a different gene, so we will be lucky if we can plug our malaria genes into an effective existing expressor system. However, we can be optimistic that we will eventually be successful. Large amounts of foreign proteins have already been made in recombinant systems. The potential yields from microbial cultures can be calculated knowing the cell density, the molecular weight of the protein, and the number of molecules produced per cell. A culture of *E. coli* growing at a density of 6×10^{13} cells/l producing 10^6 molecules of 20,000 MW per cell will yield the recombinant-coded product at about 2 g/l, equivalent to about 5% of total cell protein. Secreting systems may eventually produce more. Eukaryotic systems may be useful for proteins which are not well expressed in bacteria or which need post-translational modification, especially glycosylation, for full activity. Proteins which have already been well expressed in recombinant systems include Foot and Mouth Disease Virus antigen as a hybrid protein (17% of total cell protein in *E. coli*[23]), human leucocyte inteferon (10^6 molecules per cell in yeast[24]), and influenza hemagglutinin (10^8 molecules per cell in a mammalian SV40 system[25]).

We have not yet reached the stage where genes coding for parasite antigens, other than the variant surface glycoproteins (VSGs) of trypanosomes, have been cloned and characterised, Work on the kinetoplast DNA and VSG genes of trypanosomes have helped convince molecular biologists that parasites are respectable forms of life, capable of being studied in the laboratory. Those studies serve largely to illustrate the role of recombinant DNA as a research tool for understanding bizarre, unique, and important aspects of parasite biology. It is important, however, that more classical parasitologists understand the results and implications of such work which, in the particular case of the VSG gene studies, is relevant to many other facets of antigenic variation. Unfortunately, those studies appear to eliminate any possiblity of producing a vaccine for African trypanosomiasis, based on any immunization concept directly involving the variant antigens. On the other hand, the combination of protein and DNA approaches have identified a so far unique event in the synthesis of VSGs[26,27] which indicates yet another unusual cellular pathway in trypanosomes that could be a target for drug development. We expect the work on malaria to provide an illustration of the greatest potential application of molecular biology to human medicine: that is, to provide a unique and effective means for controlling a major disease that would not be achievable by any other route.

The application of molecular biology to parasitology offers unlimited opportunities for accelerating our investigations of the genetic, cellular, and metabolic basis of parasitism. In some cases we are optimistic that these investigations will lead to major practical advances in the detection, prevention, and treatment of parasitic diseases. Parasitologists and molecular biologists must collaborate in reformulating old problems in new terms in which they will become amenable to the molecular approach.

References

1. Watson JD, Crick FHC (1953) Molecular structure of nucleic acid. A structure for deoxyribose nucleic acid. *Nature* 171: 737–738
2. Judson HF (1979) The eighth day of creation. Simon & Schuster, New York
3. Sibatani A (1981) Molecular biology: a paradox, illusion and myth. Trends in Biochemical Sciences, 6, vi–ix
4. Kurland CG (1982) Molecular biology—an alternative view. Trends in Biochemical Sciences, 7, 46–47
5. Kohler G, Milstein C (1975) Continuous cultures of fused cells secreting antibody of predefined specificity. Nature 256: 495–497
6. Yoshida N, Nussenzweig RS, Potocnjzk P, Nussenzweig V, Aikawa M (1980) Hybridoma produces protective antibodies directed against the sporozoite stage of malaria parasite. Science 207: 71–73
7. Potocnjak P, Yoshida N, Nussenzweig RS, Nussenzweig V (1980) Monovalent fragments (Fab) of monoclonal antibodies to a sporozoite surface antigen (Pb44) protect mice against malarial infection. 151: 1504–1513

8. Holder AA, Freeman RR (1981) Immunization against blood-stage rodent malaria using purified parasite antigens. Nature 294: 361–364
9. Perrin LH, Ramirez E, Er-Hsiang L, Lambert PH (1980) *Plasmodium falciparum:* characterization of defined antigens by monoclonal antibodies. Clin Exp Immunol 41: 91–96
10. Perrin LH, Ramirez E, Lambert PH, Miescher PA (1981) Inhibition of *P. falciparum* growth in human erythrocytes by monoclonal antibodies. Nature 289: 301–303
11. Epstein N, et al (1981) Monoclonal antibodies against a specific surface determinant on malarial *(Plasmodium knowlesi)* merozoites block erythrocyte invasion. J Immunol 127: 212–217
12. Rener J, Carter R, Rosenberg Y, Miller LH (1981) Anti-gamete monoclonal antibodies synergistically block transmission of malaria by preventing fertilization in the mosquito. Proc Nat Acad Sci 77: 6797–6799
13. Watson JD, Tooze J (1981) The DNA story—a documentary history of gene cloning. W.H. Freeman & Company, San Francisco
14. Roberts RJ (1982) Restriction and modification enzymes and their recognition sequences. Nucleic Acids Res 10: 117–144
15. Orkin SH (1982) Genetic diagnosis of the fetus. Nature 296: 202–203
16. Falkow S (1981) Plasmids, transposons and gene cloning. In: Michal F (ed) Modern genetic concepts and techniques in the study of parasites. Schwabe & Co, Basel, pp 27–50
17. Lerner RA, Green N, Olson A, Shinnick T, Sutcliffe JG (1981) The development of synthetic vaccines. Hosp Pract December: 55–62
18. Racaniello VR, Baltimore D (1981) Cloned poliovirus complementary DNA is infectious in mammalian cells. Science 214: 916–919
19. Gilbert W, Villa-Komaroff L (1980) Useful proteins from recombinant bacteria. Sci Am 242: 68–82
20. Dore E, Birago C, Frontali C, Battaglia PA (1980) Kinetic complexity and repetitivity of *Plasmodium berghei* DNA. Mol Biochem Parasitol 1: 199–208
21. Borst P, Fase-Fowler F, Frasch ACC, Hoeijmakers JHJ, Weijers PJ Characterization of DNA from *Trypanosoma brucei* and related trypanosomes by restriction endonuclease digestion. Mol Biochem Parasitol 1: 221–246
22. Hoeijmakers JHJ, Borst P, Van Den Burg J, Weissman C, Cross GAM (1980) The isolation of plasmids containing DNA complementary to messenger RNA for variant surface glycoproteins of *Trypanosoma brucei.* Gene 8: 391–417
23. Kleid DG, et al (1981) Cloned viral protein vaccine for Foot-and-Mouth Disease: Responses in cattle and swine. Science 214: 1125–1128
24. Hitzeman RA, Hagie FE, Levine HL, Goedeel DV, Ammerer G, Hall BD (1981) Expression of human gene for interferon in yeast. Nature 293: 717–722
25. Gething M-J, Sambrook J (1981) Cell-surface expression of influenza haemagglutinin from a cloned DNA copy of the RNA gene. Nautre 293 620–625
26. Boothroyd JC, Cross GAM, Hoeijmakers JHJ, Borst P (1980) A variant surface glycoprotein of *Trypanosoma brucei* synthesised with a C-terminal "tail" absent from purified glycoprotein. Nature 288: 624–626
27. Holder AA, Cross GAM (1981) Glycopeptides from variant surface glycoproteins of *Trypanosoma brucei:* C-terminal location of antigenically crossreacting carbohydrate moieties. Parasitology 2 135–150

18

The Role of Membrane Research in the Future Control of Parasitic Diseases

Carlos Gitler

Membrane structures underly almost every function of the living cell. The plasma membrane delimits the cell from its environment and is therefore responsible for cellular *individuality* and *irritability*.

The exofacial surface of the plasma membrane—that in contact with the external milieu—must contain receptor-molecules which participate in the recognition of:

1. solutes that must be internalized,
2. molecules which guide the cell toward its preferred location and allow it to reside therein,
3. molecules which guide the cell along its path of differentiation,
4. molecules which represent a challenge to the cell and which elicit a cellular protective response.

The interaction of external molecules with exofacial receptors leads to transmembrane signals which couple the external events to an intracellular response occurring initially in the endofacial surface of the plasma membrane. Subsequently, such signals lead to the accretion of intracellular molecules which translate the original signal into an organized cellular response. Thus, the cell may become polarized, may organize its actomyosin networks to perform locomotion, may secrete toward a given locus, etc.

Intracellular membranes play a role in the segregation of components within the cell leading to the formation of organelles and of gradients. Membranes are responsible for creating local conditions where reactions such as electron transport may occur in nonaqueous environments and where ionic gradients can lead to the generation of electrical and chemical potentials.

This plethora of functions precludes the discussion of but a few salient aspects. They exemplify how membrane research can readily contribute knowledge which should have an immediate impact in the short-term eradication of parasitic diseases. I will first deal with the methodology available today to identify those parasitic components which lead to its individuality and irritability. Then I will discuss how from our own experience we have identified a factor that may underlie *Entamoeba histolytica* contact killing of host cells, and finally I will discuss the problem of membrane individuality and its implications in host–parasite interactions.

Methodological State of the Art

It is my opinion that methodology is available today to identify and isolate most components of the membranes of parasites which are of interest. Immunological methods are available which can be used as the basis to the characterization of the desired elements present in cell membranes.

For example, methods exist today to separate membrane lipids into distinct classes either by two-dimensional thin-layer or high-pressure liquid-chromatography. Recently, techniques have been developed to perform solid-state radioimmunoassays of lipids with great ease and sensitivity.[1] Furthermore, specific lipid-containing liposomes may be used to obtain, when injected into animals, high antibody titers to specific lipid classes. They can also be utilized to purify these antibodies.[2] Application of these techniques has lead, among other results, to the purification of a unique aminoglycophospholipid from *Entamoeba histolytica* which is responsible for the greater part of the surface-directed antibodies in patients with liver abscess and in rabbits immunized with total amoeba homogenates.[2] A unique immunogenic lipid present in the malaria parasite and in *Leishmania* has also been identified.[3] Since these lipids can be obtained in a pure form in reasonable amounts, they represent unique antigenic and immunogenic species worth pursuing for diagnostic and vaccination schedules.

In the case of membrane proteins it appears that the procedure to follow is first to develop specific antibodies to the desired protein and then to use these antibodies in order to purify further the desired molecular species. Thus, when a protein is desired that may mediate in a given parasite function, the first step would be to use the function to search for monoclonal antibodies by their interference with the function under study. Once such antibodies are identified, they may be used as immunoadsorbents to isolate the protein itself. Many interesting procedures of this type are now available. For example, formylmethionyl-peptides are known to be chemotactic signals to monocytes. Using F-met-peptides as haptens, a mono-clonal antibody was obtained. Then an antiidiotype was used to purify the chemotactic membrane receptor.[4] Monoclonal antibodies directed against other specific surface proteins can readily be obtained once a method to identify them is available. The antigen may then be obtained by the use of the antibody, taking advantage of the fact that methods to elute the antigen from immunoadsorbents are now available. Thus, either very low ionic strength buffers or 1 M MgCl have been used for this purpose.

When monoclonal antibodies cannot be obtained directly by immuni-zation of crude fractions because of antigenic dominance, alternative strat-egies are available. One of these which we have recently developed involves the separation of protein fractions in SDS-PAGE. The desired pro-tein is then eluted with detergent and mixed with detergent-containing lipids, and the detergent is removed either by dialysis or by other conven-tional techniques. The proteoliposomes thus generated can then be used to elicit an immune response in mice. We have recently shown the adju-vant qualities of liposomes: As little as 0.1 μg of the synthetic polymen TGAL trapped within a liposome injected intraperitoneally into mice resulted in an immune response equivalent to that of 100 times more of the polymer in complete Freund's adjuvant.[5] By these procedures, for example, antibodies have been obtained against a 140 kd surface protein of *Plasmodium* merozoytes which had failed to elicit the formation of monoclonal antibodies. Since this antibody against the [35]-S-methionine-labeled protein were obtained by its injection as a proteoliposome into mice, the corresponding monoclonal antibody may now be obtained. Thus even though the acrylamide gel-derived protein might not be pure, this procedure allows the elaboration of a highly specific monoclonal antibody.[6]

Many other methodological procedures such as blotting to nitrocellu-lose, followed by binding with specific lectins or antobodies can be used to identify the desired protein(s). In addition, procedures using non-pen-etrant reagents are available to label proteins present in the exofacial sur-face of the cell. In addition, proteins in contact with membrane lipids can be identified by their labeling with iodonaphthylazide.[7] In summary, it appears reasonable to think that almost any protein that may be identified can be purified by the above techniques. This is clearly a generalization, but it seems profitable to test this strategy in the various parasitic systems.

It is reasonable to think that these techniques should yield readily var-ious pure antigens to be used as possible vaccines. The problem now will reside in how to use these to induce the desired immune response. Thus for example, a local IgA-type antibody may be desirable when planning a strategy against amoeba invasion in the large intestine. The question then it how to immunize to obtain the desired isotype as a long-lasting response. Research into surface-immunoglobulin switching and its control is clearly required.

How Do Cells Kill Other Cells?

It was mentioned above that identity of a cell depends on the integrity of the plasma membrane. It is not surprising therefore that attempts of one cell to kill another in many instances respresent attempts to destroy the integrity of the plasma membrane. Since the list of compounds developed in nature for these purposes is quite large, suffice it to mention ionophoric antibiotics which collapse ion-gradients, channel-forming toxins such as bee-venom melitin, diphtheria toxin, colicins, and the antibody-complement cascade. Cells also induce changes in target cells by establishing cell–cell contact followed by damage to the target cell. *Entamoeba histolytica* has been found to kill target tissues by a contact-dependent cytolytic mechanism which is remarkably effective. Thus even at ratios of one amoeba to 50 macrophages, amoeba readily contact the cells transiently and thereafter induce cytotoxic effects followed finally by phagocytosis. Recently, we have found[8,9] that *Entamoeba* contain a remarkable material which *spontaneously* incorporates into lipid bilayers, liposomes, and cells. This protein, which we call *amoebapore,* inserts into membranes and forms ion-channels. We have developed a simple liposome assay to detect its activity[10] based on the development of a K^+-dependent valinomycin potential, which is collapsed as a function of amoebapore concentration. By the use of this assay, we have purified the amoebapore almost to homogeneity and are currently testing its effects on cells and whether antibodies could be made to neutralize it.

This example is given of how classical membrane techniques may be profitably applied to the possible elucidation of the contact-killing mechanism of virulent amoeba. As an interesting corollary, we have examined whether a similar ion-channel-forming material might be present in T lymphocyte killer cells. Preliminary results indicate that indeed such a channel-forming substance is present in these cells. Its properties are currently under study.

Maintenance of Membrane Purity

One final aspect that may be discussed in the present context involves how a cell defines the position that will be occupied by a given membrane protein. Coupled to this problem is that of how a cell maintains the identity of a given membrane. Recent studies indicate that membrane proteins and proteins to be secreted are synthesized in the endoplasmic reticulum. Secreted proteins pass through the membrane by a single peptide-mediated mechanism and are then located within the lumen of the ER. Some membrane proteins are inserted into the membrane by a similar mechanism, but their passage through the membrane is interrupted so that they remain embedded in the lipid phase by a segment of their polypeptide chain. Why this difference in behavior? It appears that there may be specific sequences of amino acids in these later proteins, which contain clustered basic amino acids which might represent stop-signals for the transmembrane passage of a given protein.[11] The question is then, how do these newly synthesized proteins find their final destination: some will be receptors at the cell surface; others will comprise vesicles that fuse with the plasma membrane to liberate their contents to the outside and thus accomplish secretion of proteins. No clear answer is as yet available. It has recently been observed, however, that the plasma membrane of many cells contains a mechanism whereby receptor-ligand complexes may be rapidly internalized. These complexes move by lateral diffusion in the cell surface until they reach areas known as coated pits. In these areas, the receptor-ligand complex is rapidly incorporated into a clathrin-coated vesicle, which then leads to the formation of an internalized vesicle containing the ligand and receptor facing toward the inside of the vesicle away from the cytoplasm. These vesicles rapidly fuse with lysosomes forming secondary vesicles where the ligand is then liberated. If the ligand is a lysosomal enzyme containing mannose-1-phosphate or mannose, its binding to a sugar-specific receptor in the outside of the cell leads to its redelivery to the lysosomal compartment. On the other hand, if the ligand is a low-density lipoprotein or an α_2 macroglobulin-proteolytic enzyme complex, binding to the specific surface receptor leads to rapid internalization of the receptor–ligand complex. The ligand upon delivery to the lysosome is rapidly degraded. However, the receptor is spared and reappears at the cell surface to be reutilized. These findings are very intriguing since they might indicate how proteins from one membrane may be delivered to a given site in the cell. Thus, the proteins present in the lumen of the ER may similarly bind to a lectin-like receptor present in the endofacial surface of the ER. Recognition could be by specific antennae sugars. The lectin–glycoprotein complex could move to a coated pit in the ER, which would form a coated vesicle containing in its lumen the glycoproteins to be secreted. These coated vesicles could then fuse with the Golgi apparatus where a similar lectin-like protein could recognize the glycoprotein after its sugars have been modified. The formation of coated vesicles could lead to the formation of a vesicle which, on fusion with the plasma membrane, could lead to the secretion of the glycoprotein. It is not known whether intrinsic proteins synthesized in the ER are moved to the Golgi and then to the plasma membrane by a similar mechanism. It seems unlikely since tunicomycin, which interferes with carbohydrate systhesis, does not affect their final destination. It is clear, however, that these proteins are selectively removed from ER to the Golgi and then to the plasma membrane. Thus receptors must be present that selectively remove them and transport them to their final destination. If the signals here are not specific sugar sequences, are they specific amino acid sequences? It is likely that in the next few years, these signals will be identified. What the above discussions possess is the very interesting possibility that each membrane will contain specific mechanism to remove proteins that are not part of that membrane and to deliver them to a given cell site or to their breakdown. Clearly, in parasites such reactions are of great interest: How are membrane proteins synthesized and delivered to their final destination? How are receptors recirculated? Answers to these questions will allow our understanding of problems such as surface antigenic modulation and surface camouflage.

This is clearly a superficial discussion of the problem, biased by personal interests. There is, however, little doubt that modern membrane concepts and techniques if applied to the problems of parasitology will in a short time contribute significantly to the possible cure of these diseases. The problem is one of economic support and a moral commitment of those who know these concepts, in order to help those individuals whose lives are hampered by these maladies.

References

1. Smolarsky M (1980) A simple immunoassay to determine the binding of antibodies of lipid antigens. J. Immunol. Methods 38:85–94
2. Calef E, Gitler C, (to be published) Identification of a unique antigenic lipid of *Entamoeba histolytica*
3. Londner et al (1982)
4. Marasco WA, Becker EL, (1982) Anti-idiotype as antibody against the formyl peptide chemotaxis receptor of the neutrophil. J. Immunol. 128:963–968
5. Lifschitz R, Gitler C, Mozes E (1981) Liposomes as immunological adjuvants in eliciting antibodies specific to the synthethic peptide (T_1G)-A—L with high frequency of site-associated idiotypic determinants. Eur J Immunol 11: 398–404
6. Hudson DE, Miller LH, Richards RL, Alvig C, Gitler C (to be published) A unique 140,000 molecular weight protein on the surface of *(Plasmodium knowlesi)* malaria merozoites. J Immunol
7. Gitler C, Bercovici T (1980) The use of lipophilic photoactivatable reagents to identify the lipid embedded domains of membrane proteins. NY Acad Sci 346: 199–211
8. Lynch E, Harris A, Rosenberg IM, Gitler C (1980) A natural protozoan-derived ionophore: a possible mechanism of lytotoxicity by *Entamoeba histolytica*. Biol Bull Woods Hole, 159: 496–497
9. Lynch E, Rosenberg IM, Gitler C (to be published) An ion-channel forming protein produced by *Entamoeba histolytica*. Embo J
10. Loew LM, Gitler C, Rosenberg IM, Lynch E (1982) Development of a fluorescent assay of ionophoric activity applied to a novel channel formed from *E. histolytica*. Biophys J 37: 143a
11. Sabatini DD, Kreibich G, Morimoto T, Adesnik M (1982) Mechanism for the incorporation of proteins in membranes and organelles. J. Cell Biol. 92: 1–22

19

Cell Biology and the Future of Parasitology

Adolfo Martínez-Palomo

"For over the past quarter of a century, no field of science has made more spectacular strides than cell biology. And over the balance of the century, no other field of science offers greater promise to increase our understanding of life and our ability to deliver quality health care." With these words Senator Edward Kennedy opened a symposium entitled "The Biological Revolution: Applications of Cell Biology to Public Welfare" held at the First International Conference on Cell Biology in 1976.[1] The symposium included a number of conferences on the application of recent advances in cell biology to current medical and environmental issues. The areas that were covered included cancer, biology of reproduction, behavior, DNA recombination and genetic disease, and environmental carcinogens, but the contributions of cell biology to the study of parasitic diseases were not mentioned. In this paper I will review some of the most important findings concerning the cell biology of parasitism, as well as speculate on future perspectives.

The Field of Action of Cell Biology

A presentation dealing with the application of cell biology to parasitology should start by defining the field of action of that discipline, one of the youngest and fastest growing branches of experimental biology, born only two decades ago. To define cell biology is neither irrelevant nor simple. In fact, most of the many textbooks on the subject carefully avoid doing so, as their contents may range from being a collection of electron micrographs to a typical monograph on biochemistry.

The term *cell biology* was originally coined in 1863 by Carnoy in his "La Biologie Cellulaire. Etude Comparé de la Cellule dans les Deux Regnes," but the successful merging of biochemistry and morphology was not accomplished until the late 1950s, exemplified by the elegant work of George Palade and Albert Claude,[2,3] The original concept of cell biology—the integration of structure, biochemistry, and function—has now been superseded by attempts to unravel the secrets of the regulatory activity of cellular constituents within the context of the overall functioning of the cell.[4]

Cell biologists have boldly dismantled the components of the complex cell machinery. Isolated and nurtured in a suitable environment, ribosomes, mitochondria, and membranes perform their stipulated tasks. For the time being, however, this cellular dissection resembles more a demolition than a disassembly, since reconstitution of the separated components still eludes the investigator.

The growth of cell biology has been so rapid that its limits have become more and more difficult to define. The great leap forward from classical cytology to cell biology was catalyzed by borrowing concepts and techniques from biochemistry, molecular biology, and cell physiology. This has occurred to such an extent, however, that the identity of cell biology is at stake. Nevertheless, it is possible and constructive to maintain the integrative view of cell biology, which is its chief value; that is, the holistic approach to the understanding of cell function attained through the use of its own specialized techniques, as well as the methodologies of other disciplines with more reductionistic perspectives.

Ultimately, the strict definition of cell biology is inconsequential, as we are not observing nature as it is, but nature exposed to our own method of inquiry. Thus, the separation of disciplines reflects an attempt to organize our ignorance rather than being a demonstration of our wisdom. This schematization, is required, however, to frame our approach to the study of cell function and structure.

One of the great achievements of cell biology, of particular relevance to parasitology, is the demonstration of the universality of cellular organelles and macromolecular assemblies. A common basis of biological organization exists not only at the molecular level, but also in the membranes, ribosomes, and mitochondria of eukaryotic aerobic cells. Parasitism can therefore be conceptualized as the interaction between different cell populations with essentially similar bioengineering patterns. Thus, the analysis of the constitutents of hosts and parasites should always be carried out with the realization that the aim is the understanding of interactions between cellular organisms.

Cell biology has demonstrated that spatial organization, motility, and intracellular traffic depend on a complex cytoskeleton, and parasites have proved to the excellent models. The explosive accumulation of data concerning the cell surface and contact-mediated phenomena have shed light on mechanisms which are crucial for the understanding of the host–parasite interplay. These determine whether the parasite confronted with an alien environment will survive, multiply, colonize, and perhaps damage its mammalian host.

The Parasite

Perhaps the greatest scientific challenge, and potentially the most useful one from the biomedical point of view, is the application of cell biology to the analysis of the differentiation and development of parasites. A few examples will illustrate this point.

Transmission of most intestinal parasitic infections is primarily carried out by encysted forms of protozoa and shelled eggs of nematodes. Interruption of the life cycle of these parasites through intervention in their developmental process could provide biological alternatives for parasite control. Information on the structure, composition, and synthesis of cell walls has mainly been acquired in fungi, whose cell walls contain chitin microfibrils embedded in a matrix of β-glucan. In these organisms it appears that biosynthetic enzymes are synthesized and packed in small vesicles called chitosomes, specific organelles that initiate chitin synthesis and eventually fuse with the plasma membrane at sites of cell wall growth.[5]

Although cellulose has long been known to be one of the main components of the cyst wall of free-living amebas,[6] essentially nothing was known about the cell walls of intestinal protozoa until microfibrils of chitin were recently demonstrated by sugar analysis and X-ray diffraction in purified cyst walls of *Entamoeba invadens*.[7]

Nematode eggshells facilitate transmission of parasites whose free-living forms show a high mortality, and are responsible for the resistance of the egg to anthelmintics. A considerable variation in eggshell structure has been found in different nematodes, but basically, shells are made of a lipid layer that constitutes the main permeability barrier, a thick chitinous layer that offers structural strength, and an outer lipoprotein vitellin layer.[8] Further information on the chemistry of cyst walls and eggshells, on their mode of secretion, and the metabolic pathways involved may disclose means for the interruption of the cell cycle, for example, through the use of inhibitors of enzymes that participate in the synthesis of chitin, which possibly could be safely used in man.

Along similar lines, much benefit might be derived from the study of the differentiation of the nematode cuticle.[9] Electron microscopy has shown the filamentous nature of this covering in microfilariae and has delineated the differentiation process that appears to result in the masking of the plasma membrane and its antigenic components, thus protecting the living microfilariae from host recognition.[10]

Variations in the surface components of pathogenic protozoa such as *Trypanosoma cruzi* during different developmental stages have been detected with electron microscopy, and subsequently have been explored using biochemical and immunochemical techniques, providing useful information concerning the immunology and pathogenesis of Chagas' disease.[11-14]

Certain areas of research may appear unglamorous by the present standards of molecular analysis but are, nevertheless, crucial for the advancement of parasitology since they can lead to an increase in the level of experimental sophistication. One example is the development of reliable techniques for the differentiation of morphologically similar species or strains of parasites, both protozoa and nematodes. During the last three years, comparison of the enzyme patterns, or zymodemes, of many isolates have suggested that the distinct clinical forms of Chagas' disease are due to differences in the etiological agent,[15] that parasites in patients afflicted with malaria are heterogeneous,[16] and that pathogenic and nonpathogenic *Entamoebae* may constitute different biological species.[17]

The genetic diversity of parasites obtained from infected patients or in cultures has to be taken into consideration in the development of therapeutic and preventive measures. Cloned parasites represent the first attempt to study the genetic composition of the isolates, which cannot be characterized by chromosome analysis due to the absence of well-defined chromosomes in the dividing nuclei of many parasitic protozoa. In the case of *Entamoeba histolytica,* a means for identifying nonpathogenic strains would settle important epidemiologic issues and would probably spare millions of people from unnecessary treatment and anxiety, as well as provide a more rational approach to the parasitological diagnosis of intestinal amebiasis.

In a few parasites such as *Toxoplasma gondii,* the availability of clones has already made possible sophisticated studies on the biochemistry of host–parasite interactions.[18] In addition, genetic recombination has provided a new means for studying the genetics of this parasite.[19]

Research on the biochemical genetics of pathogenicity could pave the way for the determination of virulence attributes and possible drug resistance of isolates. At the same time, a better characterization of surface constituents may lead to improved diagnostic and preventive measures.

The Host

Basic aspects of the host–parasite interaction may yet be unraveled by placing emphasis on the host. For example, this focus could yield information on the cellular basis of genetic resistance and susceptibility to parasitic diseases. A deeper knowledge of the mechanisms of normal resistance to microbial and parasitic infections in lining epithelia has been obtained through the exploration of the antiadherence activity of both natural and synthetic surface glycosaminoglycans in the bladder, even on molecular and ionic levels.[20]

New and important findings have also been obtained even at the histologic level, such as the demonstration of specialized epithelial cells in the intestinal mucosa at the Peyer's patches. These cells sample macromolecular particles present in the lumen of the gut and transport them to underlying lymphocytes in the first step of the afferent limb of the local immune system.[21,22]

Further probing into regional differentiation of living epithelial cells has shown that invasion of the cecal mucosa in experimental amebiasis initially takes place only at interglandular sites, where cells are continuously being sloughed off as a result of the normal renewal of the epithelium. Through these sites of lowered resistance, amebas invade the submucosa and create microulcerations that, by confluency, give rise to typical amebic ulcers.[23]

These few examples illustrate the argument in favor of further investigations into the cell structure and function of tissues that constitute the first line of defense against infections.

Switching from tissues to cells, the analysis of the cellular basis of nonspecific immunity has provided some of the most persuasive examples of the usefulness of cell biology in this area. We know at present that minutes after encountering chemoattractants, phagocytes modify their morphology and become oriented toward the chemotactic gradient. This phenomenon is initiated by the binding of chemoattractants to plasma membrane receptors, followed by alterations in transmembrane potential, changes in ionic fluxes and cyclic nucleotide levels, modifications in membrane phospholipid composition, and reorganization of microtubules and actin microfilaments.[24] These observations have permitted an understanding of the mechanism of action of agents that control inflammation such as steroids and colchicine, and could facilitate the identification of agents capable of stimulating the cellular response in sites of depressed inflammation, such as ascorbate and levamisole, which appear to improve chemotaxis in certain patients.[25]

The above-mentioned advances were made a hundred years after Metchnikoff discovered phagocytes in starfish larvae. Likewise, an initial understanding of eosinophil functions, primarily of their helminthotoxic capacity,[26] has been obtained a century after the description of the eosinophil by Ehrlich. The central role played by cell biology in these discoveries is as undeniable as is the promise of its usefulness for future exploration of the functions of these cells.

Host–Parasite Interaction

Cell biology has its clearest application to parasitology in the analysis of cellular interactions that are central to parasitism. Insight into the mechanism by which obligate intracellular protozoa find a suitable environment inside deceived professional phagocytes has already been achieved.[27] As a consequence, we know that *Trypanosoma cruzi* lyses phagolysosomes after phagocytosis,[28] *Toxoplasma* blocks lysosome-phagosome function,[29] and *Leishmania* survives within phagolysosomes,[30] Active research on the mechanism of entry of these protozoa and malarial parasites into their mammalian hosts[31] is currently being pursued. Less is known about host-parasite compatibility in extracellular protozoa and helminths.

In addition to defining the cellular basis for parasite adaptation to the host environment, the methods of cell biology may be used to study the cellular factors that determine the pathogenicity of parasites. Among these might be collagenase activity, which was recently detected in the invader *par excellence, Entamoeba histolytica.*[32]

Finally, we must accept that the eradication of parasitic diseases lies mostly out of the realm of cell biology, since they are mainly related to poverty, ignorance, and exploitation. The application of this science to some of the most important diseases of mankind, however, not only dignifies man by providing understanding of his own suffering, but also gives hope with respect to the possible cure and prevention of diseases that should not exist.

References

1. Kennedy EM (1979) Biomedical science in an expectant society. In: Weissmann G (ed) The biological revolution. Applications of cell biology to public welfare. Plenum Press, New York, pp 11–19
2. Palade GE (1975) Intracellular aspects of the process of protein synthesis. Science 189: 347–358
3. Palade GE (1971) Albert Claude and the beginnings of biological electron microscopy. Cell Biol 50: 5D–19D
4. Palade GE (1977) Greetings to the I International Congress on Cell Biology. In: Brinkley DR, Porter KR (eds) International cell biology. 1976–1977. Rockefelller University Press, New York, pp vii–x
5. Bartnicki-García S (1981) Role of chitosomes in the synthesis of fungal cell walls. In: Schlessinger D (ed) Microbiology—1981. American Society for Microbiology, Washington DC, pp 238–244
6. Tomlinson G, Jones EA (1962) Isolation of cellulose from the cyst wall of a soil amoeba. Biochim Biophys Acta 63: 194–200
7. Arroyo-Begovich A, Cárabez-Trejo A, Ruíz-Herrera J (1980) Identification of the structural components in the cyst wall of *Entamoeba invadens.* J Parasitol 66: 735–741
8. Wharton D (1980) Nematode egg-shells. Parasitology 81: 447–463
9. Bird AF (1980) The nematode cuticle and its surface. In: Nematodes as biological models. Academic Press, New York, pp 213–236
10. Martínez-Palomo A (1978) Ultrastructural characterization of the cuticle of *Onchocerca volvulus* microfilaria. J Parasitol 64: 128–136
11. De Souza W, Argüello C, Martínez-Palomo A, Trissl D, González-Robles A, Chiara E (1977) Surface charge of *Trypanosoma cruzi:* Binding of cationized ferritin and measurement of electrophoretic mobility. J Protozool 24: 411–415

12. De Souza W, Martínez-Palomo A, González-Robles A (1978) The cell surface of Trypanosoma cruzi: Cytochemistry and freeze-fracture. J Cell Sci 33: 285–289

13. Gutteridge WE (1981) Trypanosoma cruzi: Recent biochemical advances. Transactions of the Royal Society of Tropical Medicine and Hygiene 75: 484–492

14. Colli W, Andrews NW, Zingales B (1981) Surface determinants in American trypanosomes. In: Schweiger HG (ed) International cell biology 1980–1981. Springer-Verlag, Berlin, pp 401–410

15. Miles MA, Povoa MM, Prata A, Cedillos RA, De Souza AA, Macedo V (1981) Do radically dissimilar Trypanosoma cruzi strains (zymodemes) cause Venezuelan and Brazilian forms of Chagas' disease? Lancet 1: 1338–1340

16. Rosario V (1981) Cloning of naturally occuring mixed infections of malaria parasites. Science 212: 1037–1038

17. Sargeaunt PG, Williams JE, Grene JD (1978) The differentiation of invasive and non-invasive Entamoeba histolytica by isoenzyme electrophoresis. Trans R Soc Trop Med Hyg 72: 519–521

18. Pfefferkorn ER, Schwartzman JD (1981) Use of mutants to study the biochemistry of the host–parasite relationship in cultured cells infected with Toxoplasma gondii. In: Schweiger HG (ed) International cell biology 1980–1981. Springer-Verlag, Berlin, pp 411–420

19. Pfefferkorn LC, Pfefferfrom ER (1980) Toxoplasma gondii: Genetic recombination between drug resistant mutants. Exp Parasitol 50: 305–316

20. Parsons CL, Stauffer Ch, Schmidt JD (1980) Bladder-surface glycosaminoglycans: An efficient mechanism of environmental adaptation. Science 208: 605–607

21. Owen RL (1977) Sequential uptake of horseradish peroxidase by lymphoid follicle epithelium of Peyer's patches in the normal unobstructed mouse intestine: An ultrastructural study. Gastroenterology 72: 440–451

22. Keren DF (1980) Immunology and immunopathology of the gastrointestinal tract. American Society of Clinical Pathologists, Chicago

23. Mora-Galindo J, Martínez-Paloma A, González-Robles A (1983) Interacción entre Entamoeba histolytica y epitelio cecal del cobayo. Arch Invest Med (Mex) 13(suppl 3): 223–243

24. Synderman R, Goetzl EJ (1981) Molecular and cellular mechanisms of leukocyte chemotaxis. Science 213: 830–837

25. Davis JM, Gallin JI (1981) The neutrophil. In: Oppenhiem JJ, Rosentreich DL, Potter D (eds) Cellular functions in immunity and inflammation. Edward Arnold, London, pp 77–102

26. Butterworth AE, David JR (1981) Eosinophil function. N Engl J Med 304: 154–156

27. Jones TC (1981) Interactions between murine macrophages and obligate intracellular protozoa. Am J Pathol 102: 127–132

28. Nogueira N, Cohn Z (1976) Trypanosoma cruzi: Mechanism of entry and intracellular fate in mammalian cells. J Exp Med 143: 1402–1420

29. Jones TC, Hirsch JG (1972) The interaction of Toxoplasma gondii and mammalian cells. II. The absence of lysosomal fusion with phagocytic vacuoles containing living parasites. J Exp Med 136: 1173–1194

30. Chang PK, Dwyer DM (1976) Multiplication of a human parasite *(Leishmania donovani)* in phagolysosomes of hamster macrophages *in vitro.* Science 193: 678–680
31. Aikawa M, Miller LW, Rabbege JH, Epstein N (1981) Freeze-fracture study on the erythrocyte membrane during malarial parasite invasion. J Cell Biol 91: 55–62
32. Muñoz ML, Calderón J, Rojkind M (1982) The collagenase of *Entamoeba histolytica.* J Exp Med 155: 42–51

20

Immunology and
Parasitic Diseases

Peter Perlmann

Although immunology as a science has evolved from the study of infectious diseases, research in specific host defence against protozoan or helminthic infection has long been a relatively neglected area. More recently, however, this situation has begun to change, and the immunology of parasitism is becoming more attractive. Several facts account for this. In part it reflects a growing appreciation of the actual need to fight those diseases which constitute the major health problem of the world, affecting the physical and social well-being of hundreds of millions of human beings. On the other hand, another important reason for this beginning shift of emphasis probably must be sought in the dramatic advances presently being made in many central areas of immunology itself. These advances are rapidly deepening our understanding of how the immune response is regulated and how it can be manipulated to become protective. Tissue damaging and protective immune effector mechanisms are being elucidated at the cellular and molecular level, and new tools for improved immunodiagnosis of infectious agents are being developed at a rapid pace. For these reasons, immunologists are much better equipped today than only a few years ago to contribute to the elucidation of the complicated interplay between the vertebrate host and the protozoan or multicellular parasite. Hence, it is not surpirsing that application of new immunological knowledge to the field of parasitism is seen by many as a challenge—a challenge which may be expected to render beneficial results both in the further advancement of immunology as a science and, above all, in the fight against parasitic disease.

In this brief survey I will illustrate with a few recent examples the impact on parasitology of what is sometimes called the "new" immunology. I will also point out some of the directions where important results may be expected to be forthcoming within the next few years. I will restrict myself to discuss two main areas, namely (1) various uses of monoclonal antibodies, and (2) recent advances in the field of immune regulation, relevant in this context. I will not discuss immune effector mechanisms believed to be important in the vertebrate's defence against multicellular parasites. Here, Dr. Capron gives an excellent example in Chapter 8 of this volume by discussing the roles of anaphylactic IgG and IgE antibodies in triggering cellular defence mechanisms in schistosomiasis (see also references 1,2).

Monoclonal Antibodies

The introduction of the hybridoma technology by Köhler and Milstein[3] for the production of monoclonal antibodies has, within a few years, profoundly affected immunology both in its basic aspects and in its applications. In contrast to what is achieved by conventional immunisation procedures, cloned B-cell hybridomas produce homogeneous antibody reagents, each strictly specific for a single antigenic determinant (epitope) on the immunogenic macromolecule, which usually has several different such structures. In combination with relatively simple selection and cloning procedures, the investigator can therefore pick up those hybridomas which produce the reagent of proper specificity and other molecular properties best suited to his particular purpose. Since cloned hybridomas can be kept in a productive phase over long periods of time, access to homogenenous antibody of given specificity may be practically unlimited.

One important area of application of monoclonal antibodies in parasitology concerns their use as diagnostic tools. That the properties mentioned above make monoclonal antibodies superior reagents for this purpose is witnessed by a steadily increasing number of publications in the current literature. An original example that well illustrates the extraordinary usefulness of monoclonal antibodies in this context is the "4-i" assay recently described by Potocnjak et al.[4] Here, 4-i stands for "inhibition of idiotype anti-idiotype interaction." These authors have previously reported on the production of a monoclonal mouse antibody directed against the major surface coat protein (Pb44) of sporozoites of the murine malaria parasite, *Plasmodium berghei*. This antibody has also been shown to protect mice against *P. berghei* infection. By injecting mice with this antibody, the authors produced a second monoclonal antibody directed against the idiotype of the anti-Pb44 immunoglobulin used in immunogen. This anti-idiotypic antibody reacted with a structure closely associated with the antigen binding site of the anti-Pb44 antibody. The inhibition of the interaction of these two antibodies by *P. berghei* sporozoites constituted the basis of a radioimmunoassay allowing quantitative determination of small amounts of the relevant Pb44-antigen even when present in a crude extract, for example, in salivary gland extracts of individual *P. berghei* infected mosquitoes. Since production and mode of selection of monoclonal anti-idiotype antibodies has been worked out for a variety of antigen–antibody systems, the principle of this assay is of general applicability. As in other monoclonal-antibody-based assays, the reagents are homogeneous, are obtainable in unlimited amounts, and do not require purified antigen. In contrast to other radiometric or enzyme immunoassays, antigen analysis by the "4-i" assay requires only a single epitope. This makes it particularly attractive when the antigen is a small polypeptide or fragment, or is weakly immunogenic.[4]

Notwithstanding its usefulness for diagnostic purposes, the hybridoma technology has its most spectacular impact on current progress in the characterisation and isolation of parasite antigens. This is particularly important for those parasites where humoral immunity plays a significant role in host protection and where, consequently, vaccination is considered a primary goal for achieving disease control. Malaria may be taken as a case in point. While there is evidence that antibodies to both sexual and asexual stages of the parasite are protective in both experimental and human malaria,[5] knowledge of the specific immunogens involved has been scarce. Although it has been possible to produce some vaccines using crude plasmodial extracts,[6] there is general agreement that well-defined antigens will have to be selected in order to design efficient vaccination programs. The advent of monoclonal antibodies has dramatically facilitated the search for such antigens. The *P. berghei* sporozoite coat protein, a prime candidate for a vaccine of the rodent host, has already been mentioned. Here, certain monoclonal antibodies have given valuable information about those fine structures of the parasite protein which appear to be involved in its interaction with host cells and which, therefore, may constitute the basis for protective immunity.[7] Analogous polypeptides of similar properties have been found in sporozoites of simian and human *Plasmodium* species.

Several antigens specific for the erythrocytic stages of both rodent, simian, and human plasmodia have also been characterised by means of monoclonal antibodies. In several instances, evidence for a role in protective immunity of these polypeptides has been forthcoming. Thus, immunisation of mice with schizont-derived antigen, isolated by affinity chromatography on unsolubilised monoclonal antibody, has been reported to give efficient protection against mouse malaria species *P. yoelii* infection.[8] For simian and human *Plasmodium* species, a possible role in protective immunity of certain schizont or merozoite antigens is indicated by the antibody-induced inhibition of erythrocyte invasion or parasite replication *in vitro*.[9]

A full account of this active area of research will be found in two recent reports published by the WHO.[10,11] Although it is obvious that many problems have yet to be solved before antimalaria vaccines will be available for clinical use on a large scale, reading these reports makes it clear that progress in this field has been much faster than anticipated only two or three years ago.

The exquisite specificity and homogeneity of monoclonal antibodies also makes them excellent new tools for the study of many other problems of parasitic diseases. The immunopathology of Chagas' disease, resulting from infection with *Trypanosoma cruzi,* may be taken as an example. As is well known, this disease is characterised by extensive tissue degenerations, comprising both cardiac muscle and neurons in the peripheral nervous system as well as in the brain. It has been suspected for quite some time that autoimmune processes might play a role in the causation or, at least, the perpetuation of these tissue degenerations. Some indirect evidence also supports this notion by suggesting that mammalian cardiac muscle and neurons might share antigens with *T. cruzi.* More direct evidence for this has recently come forth showing that one out of 11 monoclonal antibodies raised against rat dorsal root ganglia also recognised a surface antigen on *T. cruzi* amastigotes.[12] This cross reaction was specific for *T. cruzi.* Moreover, immunocytochemical studies indicate that this antibody reacts both with cardiac muscle (but not with skeletal muscle) and with those classes of neurons known to degenerate in Chagas' disease. Obviously, monoclonal antibodies will provide new and superior probes for the elucidation of pathogenic processes underlying one of the most serious parasitic diseases, affecting millions of people in Latin America.

In all published reports on the application of the hybridoma technology to the study of parasitism, the monoclonal antibodies used were of mouse or rat origin. In the study of human disease, the use of monoclonal antibodies of human origin will add yet another dimension to this field. Thus, such reagents will allow a more direct investigation of those antigenic structures which give rise to protective or perhaps harmful immune responses in the patient. They will also permit the study of disease-associated idiotypes and the production of anti-idiotypes (see below) and hopefully provide suitable reagents for passive immunisation. Technically, the production of human monoclonal antibodies is not more complicated than that of rodent monoclonals. Human myeloma or lymphoma cell lines suitable for hybridoma formation are already available. An additional technique, applicable only to the human system, involves transformation of specific B cells by Epstein-Barr virus into antibody-producing cell lines.[13] These techniques can also be used in combination. Both require either selection of the relevant B cells (from acutely or chronically ill patients or from immune individuals) by adsorption to unsolubilized antigen or activation by exposure to soluble antigen *in vitro.*

Regulation of the Immune Response

Research in this central area of immunology aims at an understanding of the cellular, genetic, and molecular mechanisms regulating the immune response. Clearly, this understanding is necessary for a full appreciation of why immunity is easily built up and remains stable in some infections while it may rapidly wane or be suppressed in others. Again, a relevant objective of research in this field is to improve our possibilities to manipulate the immune response, for instance, to achieve a stable acquired immunity against a parasite infection.

The immune system may be viewed as a complicated network of cellular and humoral interactions, involving the immunoglobulins, a large variety of functionally distinct lymphocytes, and their soluble products as well as several types of accessory cells. When antigen is introduced, for example, by infection or vaccination, the balance of the system is disrupted. Whatever the outcome of this disruption—be it immunity or a state of nonresponsiveness—it is always the net result of a large number of inducing, amplifying, and suppressive interactions between the various factors making up the system. Knowledge of the cellular and molecular basis of these interactions has made conspicuous advances in the past few years. Some of the implications of this, relevant in the present context, will be discussed below.

According to the hypothesis put forward by Jerne almost ten years ago,[14] one of the important regulatory networks of the immune system comprises idiotype–anti-idiotype interactions. As already stated, the idiotypes of antibodies (as well as of certain antigen binding T-cell receptors) are immunogenic structures of the variable region of molecules, frequently closely associated with their antigen binding sites. In accordance with this, the anti-idiotype is an antibody recognising a given idiotype. According to Jerne's hypothesis, formation of antibodies against foreign antigens will also result in the induction of auto-anti-idiotypic antibodies, and the balance between idiotypes and anti-idiotypes will profoundly affect the course of an immune response. Numerous experimental findings confirming various aspects of the Jerne theory have been published over the past eight years.

One of the predictions of the theory is that certain anti-idiotypic antibodies, constituting the "internal image" of a foreign antigen, may induce or maintain antibody (= idiotype) formation even in the absence of the foreign antigen recognised by that antibody. These predictions, previously investigated in a number of experimental models, have recently been tested in a parasite system as well.[15] These workers succeeded in immunising mice against African trypanosomiasis by injecting them with mouse anti-idiotype antibodies. These were raised against monoclonal antibodies previously shown to protect mice against *T. rhodesiense* infection. In several instances, anti-idiotype vaccination induced complete or partial protection in the immunised mice. Moreover, in those animals that displayed immunity, this was also associated with an early induction of elevated serum levels of the relevant idiotype.

Interference with the idiotypic network by anti-idiotype vaccination is not a trivial procedure since the outcome depends on a large number of variables, some of which are not as yet fully understood. Nevertheless, the experiments by Sacks et al.[15] demonstrate clearly that anti-idiotype vaccinations may become a feasible approach, particularly in those instances when a given parasite antigen may be difficult to come by in sufficient quantities. Be that as it may, since idiotype–anti-idiotype interactions are important regulators of the immune response at both humoral and cellular levels, further exploration of these reactions in parasitic infections will be of great significance.

The key cells in the regulation of the immune response are the T lymphocytes. Regardless of whether immune protection against a parasite is primarily cell-mediated as in leishmaniasis or antibody-mediated as in certain phases of malaria or schistosomiasis, the thymus-dependent lymphocyte system plays a decisive regulatory role in almost all instances. The lymphocytes of the peripheral lymphoid system are a highly hetergeneous family of cells, differing both in differentiation lineage, state of maturation, and antigen recognition. For the T-cell system, this heterogeneity comprises a considerable number of functionally distinct cellular subsets, includ ing inducer, amplifier, helper, and suppressor cells as well as cytotoxic effector cells. Since functionally distinct T cells also differ in their surface structures, alloantibodies (in mice)[16] or monoclonal antibodies (in man),[17] directed to various differentiations or maturation antigens, have provided valuable tools for both separation and functional studies of these lymphocytes. The results of recent investigations along these lines on the T-dependent regulation of antibody formation in experimental malaria[18] offer a good illustration of the potential value of this approach.

Soluble lymphocyte (and monocyte/macrophage) derived maromolecules have been shown to be instrumental in T-cell dependent regulation of the immune response. Although it has been known for over ten years that such "factors," usually called lymphokines, play an important role in lymphocyte and macrophage activation, it is only recently that several of them have been purified and characterised. Some of these factors regulate T- or B-cell maturation and/or proliferation in a nonspecific manner [e.g., interleukin 2 (IL2), T-cell replacing factor (TRF), various B-cell maturation and replication factors). Others are antigen (or anti-idiotype) specific and MHC-restricted (e.g., antigen specific helper or suppressor factors), that is, they require genetic compatibility between the factor and the cell whose activity it regulates. Based on these findings, a coherent picture of both T-and B-cell activation is rapidly emerging.[19,20] Of major importance for the development of this knowledge has been the recent design of methods allowing long-term culture and cloning of hybridoma formation of T-1 cell lines with preserved antigen specificity and function.[21] Such cloned T-cell lines are of great potential value in the dissection of the various specific and nonspecific events resulting either in immunity or in suppression.

An example of the application of this new knowledge to a parasite infection is provided by the work of Louis et al.[22] These workers studied T-cell function in mice infected with *Leishmania tropica,* an infection leading to the development of cutaneous lesions similar to those observed in human leishmaniasis. As in the human system, healing of the cutaneous lesions in this animal model is thought to reflect a thymus-dependent T-cell mediated immunity. By antigen stimulation *in vitro* and culture in IL-2 containing medium, these workers succeeded in developing several antigen-specific T-cell clones, some displaying helper activity on antibody production by B cells and transferring parasite-specific delayed hypersensitivity, others activating parasitised macrophages, resulting in parasite destruction.

Since antigen-specific human T cells can also be cloned and investigated by various *in vitro* methods, this new technology for studying immune regulation is not restricted to animal models.[23] In addition to its great conceptual potential, it has considerable practical implications. Culturing of T-cell clones allows biochemical and genetic characterisation of the various lymphokines and monokines involved in immune regulation. This will soon permit production on a large scale, be it by T-cell culture or recombinant DNA technology, of structurally and functionally well-defined mediators. This is already under way with the interferons belonging to the same category of factors. The clinical use of lymphokines, applied with or without antigen, will provide new means of affecting the course of an immune response in human disease, including diseases caused by parasite infection.

References

1. Capron A, Dessaint JP, Capron M, Joseph M, Torpier G (1982) Effector mechanisms of immunity to schistosomes and thier regulation. *Immunol Rev* 61: 41–66
2. Butterworth AE, Taylor DW, Veith MC, Vadas MA, Dessein A, Sturrock RF, Wells E (1982) Studies on the mechanisms of immunity in human schistosomias. Immunol Rev 61: 5–39
3. Köhler G, Milstein C (1975) Continuous cultures of fused cells secreting antibody of predefined specifity. Nature 256: 495–97
4. Potocnjak P, Zavala F, Nussenzweig A, Nussenzweig V (1982) Inhibition of idiotype–anti-idiotype interaction for detection of a parasite antigen: A new immunnoassay. Science 215: 1637–39
5. Cohen S, Mitchell GH (1978) Prospects for immunisation against malaria. In: Arber W, Henle W, Hofschneider P, et al (eds), Current topics in microbiology and immunology, Springer Verlag, Berlin, Heidelberg, New York: 80: 97–137
6. Taylor DW, Siddiqui WA (1982) Recent advances in malarial immunity. Ann Rev Med 33: 69–96
7. Hollingdale MR, Zavala F, Nussenzweig RS, Nussenzweig V (1982) Antibodies to the protective antigen of *Plasmodium berghei* sporozoites prevent entry into cultured cells. J Immunol 128: 1929–1930

8. Holder AA, Freeman RR (1981) Immunization against blood-stage rodent malaria using purified parasite antigens. Nature 294: 361–364
9. Perrin LH, Dayal R (1982) Immunity to asexual erythrocytic stages of *Plasmodium falciparum:* Role of defined antigens in the humoral response. Immunol Rev 61: 245–269.
10. Report of the fourth meeting of the Scientific Working Group on the immunology of malaria: the antigenic structure and related biology of *Plasmodia*. Geneva, 14–17 October 1980. *TDR/IMMAL-SWG(4)/80.3*
11. Report of the fifth meeting of the Scientific Working Group on the immunology of malaria: the identification, structure and function of malarial antigens. Geneva, 8–10 March 1982 (to be published)
12. Wood JN, Hudson L, Jessel TM, Jamamoto M (1982) A monoclonal antibody defining antigenic determinants on subpopulations of mammalian neurones and *Trypanosoma cruzi* parasites. Nature 296: 34–38
13. Steinitz M (1981) Human monoclonal antibodies produced by Eptein–Barr virus immortalized cell lines. In:Hämmerling GJ, Hämmerling U, Kearney JF (eds) Monoclonal antibodies and T-cell hybridomas. Elsevier/North Holland, Biomedical Press, Amsterdam, New York, Oxford, pp 447–452
14. Jerne NK (1974) Towards a network theory of the immune system. Ann Immunol (Paris) 125: 373–389
15. Sacks DL, Esser KM, Sher A (1982) Immunization of mice against African trypanosomiasis using anti-idiotypic antibodies. J Exp Med 155: 1108–1119
16. McKenzie IFC, Potter T (1979) Murine lymphocyte surface antigens. Adv Immunol 27: 179–338
17. Haynes BF (1981) Human T-lymphocyte antigens as defined by monoclonal antibodies. Immunol Rev 57: 127–161
18. Jayawardena AN (1981) Immune responses in malaria. In: Mansfield JM (ed) Parasite diseases, vol. 1. Marcel Dekker Inc., New York and Basel, pp 85–136
19. Smith KA, Ruscetti FW (1981) T-cell growth factor and the culture of cloned functional T-cells. Adv Immunol 31: 137–175
20. Möller G (ed) (1979) Activation of antibody synthesis in human B-lymphocytes. Immunol Rev 45
21. Möller G (ed) (1981) T-cell clones. Immunol Rev 55
22. Louis JA, Zubler RH, Coutinho SC, Lima G, Behin R, Mauel J, Engers HD (1982) The in vitro generation and functional analysis of murine T-cell populations and clones specific for a protozoan parasite, *Leishmania tropica*. Immunol Rev 61: 215–243
23. Callard RE, Smith CM, Beverley PC (1982) Phenotype of human T helper and suppressor cells in an *in vitro* specific antibody response. Eur J Immunol 12: 232–236

21

The Biology of Parasitism
(COURSE AT WOODS HOLE, PART II)
The Future of Immunology
(AS IT PERTAINS TO PARASITOLOGY, PART I)

John R. David

First I shall describe in more detail the Biology of Parasitism course which has been running each summer since 1980 at Woods Hole, Massachusetts, and, second, introduce the topic, "The Future of Immunology as It Pertains to Parasitology."

The Biology of Parasitism Course at Woods Hole

The Biology of Parasitism course had its third session during the summer of 1982. The thinking that led to the initiation of this course and a description of the planning for it have been presented eloquently by Paul R. Gross at the previous meeting on the "Current Status and Future of Parasitology" that was held in New Orleans in October 1980. Thus, I will refer you to his paper and not repeat this here. Instead, what follows are details of the course itself, the instructors and the students who participated, focusing on what went on in the lectures and especially in the laboratory. The main object of the course was to bring together the new concepts and tools of immunology, molecular biology, and membrane biochemistry to problems of parasitology. A further aim was to attract scientists to this area.

The Biology of Parasitism course runs for ten weeks starting in the middle of June. Members of the faculty who participated in the laboratory in 1980 or 1981 are listed in Table 21-1. This faculty made major contributions to the design of the course and especially to the work that was carried out in the laboratory during the ten-week periods. Lectures are given for one and one-half hours every morning, six days a week, for the first six weeks. In addition, there is an evening lecture once a week throughout the course, and time is also made available for discussions, including informal talks by the students presenting their previous work back home.

TABLE 21-1. Course faculty

Director:	John David	Harvard Medical School
Assistant:	Roberta David	Harvard Medical School
Instructors:	Richard Carter	National Institutes of Health
	Eli Chernin	Harvard School of Public Health
	George Cross	Wellcome Research Laboratories
	Peter David	Harvard Medical School
	Carlos Gitler	Weizmann Science Institute
	Marcel Hommel	Harvard Medical School
	Diane McMahon Pratt	Harvard Medical School
	Louis Miller	National Institutes of Health
	Miercio Pereira	Tufts New England Medical Center
	Willy Piessens	Harvard Medical School
	Dyann Wirth	Harvard University

Alan Sher from the NIH and Alain Dessein, Donald Harn, and Gina Moser from Harvard Medical School were also involved in the laboratory. Above is for 1980 and 1981.

The list of lectures for the summer of 1981 are given in Tables 21-2 and 21-3. Since a Gordon Conference on the Immunology and Biochemistry of Parasites was held in New Hampshire in August 1981, we were fortunately able to entice 13 of the participants to spend a day in Woods Hole to give a Mini Gordon Conference, the program of which is given on Table 21-4.

Sixteen students participate each year; the number of students is governed by the size of the main laboratory. Of the 32 students participating during the first two years, 7 had an M.D. degree, 11 had a Ph.D., 13 were graduate students, and one was a college student. They had a very varied background in biology. Thirteen students came from overseas, and 19 were from various institutions in the United States (see Tables 21-5 and 21-6).

The heart of the course is the laboratory work. A large old laboratory in the Lillie Building at the Marine Biological Laboratory was completely gutted and redesigned to provide laboratory benches for 16 students. Six small laboratories which open into the main laboratory and two more laboratories adjacent were redesigned for work by the instructors, as well as for specialty rooms. These included rooms with biohazard hoods and incubators for tissue culture, rooms for containment and for radioactivity, and space for infected snails. During the first year, one room was devoted to a fluorescent-activated cell sorter on load to the course. A small insectary was built to house the complete life cycle of mosquitoes, and a small animal house with special screening was set up in another building to house infected animals, including mice, guinea pigs, rats, birds, hamsters, rabbits, and chickens.

The instructors played a major role in designing the work in the laboratory. Many of the experiments had not been carried out before. The stu-

TABLE 21-2. Biology of parasitism—1981: Morning Lectures 8:30–10:00

June 15 Mon.	Louis Miller—Malaria I
16 Tues.	George Cross—Antigenic Variants in Trypanosomes I
17 Wed.	Louis Miller—Malaria II
18 Thurs.	Bryan Roberts—Principles of Recombinant DNA
19 Fri.	George Cross—Trypanosomes and Their Genes II
20 Sat.	Carlos Gitler—Membrane I
22 Mon.	Carlos Gitler—Surface Labelling—Membrane II
23 Tues.	Carlos Gitler—Membrane III
24 Wed.	Larry Simpson—DNA of Kinetoplasts
24 Wed.	Andre Spielman—Vectors
26 Fri.	Carlos Gitler—Amoeba
27 Sat.	Philip Marsden— *T. cruzi*
29 Mon.	Philip Marsden—Leishmania
30 Tues.	George Nelson—Schistosomiasis
July 1 Wed.	John Caulfield—Schistosomiasis: Parasite–Cell Interactions—EM
2 Thurs.	George Nelson—Zoonosis
3 Fri.	George Nelson—Onchocerciasis
4 Sat.	John David—Immunology—Introduction
6 Mon.	Richard Gershon—T Cells and Immune Regulation I
7 Tues.	Richard Gershon—T Cells and Immune Regulation II
8 Wed.	Philip Askenase—IgE, Mast Cells and Mediators
9 Thurs.	Philip Askenase—Cutaneous Basophil Hypersensitivity, Ticks
10 Fri.	Douglas Fearon—Complement, Classical
11 Sat.	Douglas Fearon—Complement, Classical
13 Mon.	John David—Eosinophils as Effector Cells
14 Tues.	John David—Cell Mediated Immunity
15 Wed.	Dyann Wirth—Recombinant DNA
16 Thurs.	Manfred Karnovsky—Oxygen Products
17 Fri.	Ruth Nussenzweig—Immunology of Malaria Sporozoites
18 Sat.	Willy Piessens—Immunology of Filariasis
20 Mon.	Diane McMahon Pratt—Immunology of Leishmania
21 Tues.	Miercio Pereira— *T. cruzi,* Surface, Lectins
22 Wed.	Elmer Pfefferkorn—Genetics of Toxoplasma
23 Thurs.	Mary Rifkin—Trypanosome Plasma Membrane: Interaction with Host Proteins
24 Fri.	Fotis Kafatos—Recombinant DNA Studies on the Structure, Evolution, and Developmental Expression of Structural Gene Families

dents worked in groups of four partly due to the nature of the experiments involved and partly limited by the cost involved for these experiments.

During the ten weeks of the first year's session, the students did the following:

1. They produced monoclonal antibodies to *Leishmania enriettii* and demonstrated that several of these reacted with tubulin.
2. They carried out studies on the genomic cloning of *L. enrietti* with the aim of determining if they could obtain expression of leishmania

TABLE 21-3. Biology of Parasitism—1981: Evening Lectures 7:30–8:30

June	16 Tues.	Fred Bang—Biology of Parasites of Some Invertebrates
	18 Thurs.	Class Presentation
	22 Mon.	Robert May—The Population Biology of Parasitic Infections
	30 Tues.	Irwin Sherman—Biochemistry of Malaria
July	7 Tues.	Alan Sher—Adaptation of Parasites to the Immune Response
	14 Tues.	Kenneth Warren—Immunopathology of Schistosomiasis
	22 Wed.	Carlos Gitler—Amoeba
	28 Tues.	Gerald Keusch—Viruses and Diarrheal Diseases in Tropical Medicine
Aug.	4 Tues.	Lawrence Lichtenstein—Mechanisms of Allergy
	11 Tues.	Albert Vincent—Morphology and Biology of Filaria

Lecturers in 1980 also included: Daniel Brooks, Harvey Cantor, Daniel Goodenough, Richard Guerrant, Adel Mahmoud, Edward Michelson, William Trager, Emil Unanue, Byron Waksman, and Thomas Weller.

TABLE 21-4. MBL Course on "Biology of Parasitism": "Mini Gordon Conference" on Immunology and Biochemistry of Parasitism

Speaker	Subject
Part I	
Piet Borst (University of Amsterdam)	Rearrangement and expression of VSG genes in *T. Brucei*
Paul Englund (Johns Hopkins)	Structure and replication of kinetoplast DNA
David Snary (Wellcome—Kent)	Purification of an immunogen from *Trypanosoma cruzi*
Anthony Cerami (Rockefeller University)	Trypanocidal effects of drug-induced oxidation
John Eaton (University of Minnesota)	Plasmodistatic effects of oxidants and of chloroquin
Anthony Allison (Immunologie Centre—Marseille—Luminy)	Mechanisms of cell-mediated immunity against *Thelieria pava*
Part II	
James Howard (Wellcome—Kent)	Immunoregulatory mechanisms in experimental murine leishmaniasis.
Diane McLaren (MRC—Mill Hill)	Immune evasion during schistosome development
Ron Smithers (MRC—Mill Hill)	Pre-lung and post-lung stages of immunity against *S. mansoni*
Monique Capron (Institute Pasteur—Lille)	Mechanisms of immunity against schistosomes *in vitro*
Andre Capron (Institute Pasteur—Lille)	Schistosomiasis: Parasite factors involved in host-interface regulation
Dan Colley (Vanderbilt University)	Cellular regulatory mechanisms underlying the modulation of schistosome egg granulomas
Brigit Ogilvie (Wellcome)	A novel mechanism of immunity to intestinal helminths

TABLE 21-5. Biology of parasitism
students from overseas

Kenya	3
Mexico	2
Philippines	1
Egypt	1
Sri Lanka	1
Brazil	1
Israel	1
Italy	1
Sweden	1
England	1
Total	13

TABLE 21-6. Biology of parasitism students from
United States

University of California	2
Massachusetts Institute of Technology	2
Harvard University	2
Johns Hopkins University	2
Colorado State	1
University of Texas	1
Harvard Medical School	1
University of Alabama	1
Spelman College	1
Albert Einstein College of Medicine	1
Boston University	1
New York University	1
New England Biological Laboratories	1
Cold Spring Harbor Laboratories	1
University of Cincinnati	1
Total	19

antigens. Thus, they isolated DNA from the protozoa and plasmids, and following cutting by restriction enzymes, incorporated the genomic DNA fragments into plasmids and transfected *E. coli*. Five clones reacted with a radiolabeled antibody against leishmania, but, unfortunately, these clones were not recovered.

3. They isolated messenger RNA from malaria, trypanosomes, leishmania, and ameba and translated parasite polypeptides *in vitro*.
4. Using a labeled probe for the tubulin gene of *Drosophilia* that had been given to us, they demonstrated, by hybridization, the presence of the tubulin gene in the DNA from trypanosomes, leishmania, and microfilaria, but not from *Entamoeba histolytica*.

5. They studied protective immunity and immune evasion of schisto-somes in mice that had been previously immunized with irradiated cercaria and examined the difference between the immunity to early skin schistosomula and lung schistosomula using quantitative tech-niques involving the recovery of larvae from the lung after five days and perfusion of the adult worms from mice after five weeks.

6. They demonstrated the *in vitro* killing of schistosomula by antibody and eosinophils and obtained human eosinophil preparations of 95% purity.

7. They studied the uptake of mouse major histocompatibility host anti-gens by lung schistosomula using immunofluorescence technique.

8. They learned to obtain and purify large number of *T. brucei* from rats and studied antigenic variation using immunofluorescence.

9. They labeled the surface membranes of malaria, trypanosomes, leish-mania, ameba, filaria, and starfish amebocytes with a variety of surface probes inlcuding iodogen, lactoperoxidase, diazotized sulfanilic acid, iodonaphthyl azide (the latter labels components that are within the lipid bilayer), and analyzed the products on gels by autoradiography.

10. They attempted to study some surface parasite antigens using the flu-orescent-activated cell sorter.

11. They studied the binding of several lectins to filaria and to eggs from *Schistosoma mansoni* and *Schistosoma haematobium* and found spe-cific binding patterns for each, which could subsequently be used for identification and for separation of surface components by affinity chromatography.

12. One student demonstrated the presence of a natural ionophore in extracts of *E. histolytica* which may play a role in cell cytotoxicity pro-duced by this protozoa.

13. One student demonstrated the presence of several protozoan enzymes having substrate specificity that differed from their mammalian coun-terparts and, thus, could be used in the development of new antipro-tozoan drugs.

14. They carried out studies on the motility and antigen capping of *E. his-tolytica*. One student developed a solid phase radioimmunoassay for antibodies to this organism.

15. They followed the complete life cycle of *Aedes egypti* mosquitoes, of malaria using *Plasmodium gallinaceum,* and of *Schistosoma mansoni.*

16. They studied the physiology of exflagellation of gametocytes of *Plas-modium gallinaceum,* as well as the immunology of anti-gamete antibodies.

Every student participated in the general laboratory experiments, and most also carried out additional individual projects.

In the second year, the structure of the course was similar to the first

year with a few variations and with considerable work on *Plasmodium falciparum*. This time they produced monoclonal antibodies to microfilaria
of *Brugia malayi*. They carried out genomic cloning of DNA from *Leishmania enriettii* and obtained several clones of bacteria which contained
the leishmania gene for tubulin. In studies on antigenic variation, they
isolated DNA from *Trypanosoma brucei,* and, using Southern blot technique, demonstrated the expression-linked copy of a gene for a variant
surface antigen which had not been demonstrated previously. They carried out studies on *Trypanosoma rangeli,* a nonpathogenic South American trypanosome, and showed that it bound lectins different from those
previously shown to bind to *Trypanosoma cruzi*. Further, the production
of a neuraminidase by *T. rangeli* was demonstrated for the first time.

With malaria, the students mastered the culture of *Plasmodium falciparum in vitro* and went on to carry out a number of comparative studies
with knob-positive and knob-negative clones. These included studies
using metabolic labeling, immunoprecipitation and gel analysis, fluorescent studies using infectious serum, and demonstration of the difference
in the ability of knob-positive and knob-negative clones to bind to mela-
noma cells and human monocytes. A series of morphologic studies using
video-enhanced interference phase microscopy were also carried out.
With leishmania, the students isolated and cultured amastigotes and promastigotes, compared different strains by isoenzyme analysis and kinetoplast restriction enzyme patterns, and performed cross-hybridization
studies.

The intellectual excitement and enthusiasm generated by the interactions between faculty and students has led to a plethora of new ideas that
have already had a fertilizing influence in home institutions all over the
world. This course has been directly responsible for new collaborations
and new research that would not have existed without it, for a number of
gifted new workers coming into the field, and for establishing a feeling of
warm and generous communication and exchange of ideas. These accom-
plishments, brought about in two short years, more than justify the hopes
of those who planned and supported this course at the outset.

The course was supported by major grants from The Edna McConnell
Clark Foundation and The Rockefeller Foundation. Starting in the second
year, further support was given by the Tropical Disease Research Program
of the World Health Organization and from the Wellcome Trust.

The Future of Immunology
as it Pertains to Parasitology

The present explosion of new knowledge in the field of immunology
should have a dramatic impact on parasitology in the future. I shall briefly
outline some areas which appear to be especially relevant to parasitic diseases, and Peter Perlman will expand on some of these subsequently.

Immunoregulation

One of the most active areas in immunology is concerned with how the immune system works and how it is regulated. The unraveling of the complex network of interactions between many subsets of T lymphocytes which can augment or suppress the immune response is especially pertinent to understanding how a parasite evades the immune response by tampering with this regulation, and how this might be overcome. Similarly, the new insights into the manner by which different antigens are processed and presented to lymphocytes and how the network of idiotypic and anti-idiotypic antibodies may regulate the immune system are also pertinent.

In the future, it should be possible to select antigens that do not cause suppression or to inhibit the suppressive response in other ways. Adjuvants should become available that will focus the immune response onto specific effector pathways, selecting certain T-lymphocyte subsets over others, or that turn on the production of selected classes of antibodies that are protective and not those that may block protective reactions.

Knowledge of how various parasites evade the immune response and what immune effector mechanisms exist against them should be especially useful in planning an effective immunology strategy to combat the parasites.

Further studies on the effects that different parasites have on the immunologic responses of humans should provide information crucial for development of effective vaccines in some of these diseases. In the past, inability to demonstrate a strong protective immune response to some parasites has led many to think that vaccines would not be effective. Further understanding of the mechanisms of immunoregulation has led to a new concept, namely, that it should be possible to produce protective immunity that is better than that found in nature.

Mechanisms of Immune Evasion

As mentioned above, parasites have developed some most ingenious ways of evading the immune response, such as mounting effective suppressive immune responses (the stimulation of suppressor cells and blocking antibodies), varying the antigen on their surfaces, shedding surface antigens, producing enzymes that destroy antibodies, masking themselves in host antigens, developing membranes of surface armor resistant to immune attack, and so on. Some escape the immune system by entering host cells, at the same time escaping intracellular microbicidal defenses. If we can understand how the parasites do this, we should be able to design effective ways to overcome these evasive reactions.

Effective Mechanisms of the Host

Despite their ability to evade the immune response, many parasites, at certain stages of their life cycle in their mammalian host, are susceptible to the immune system. If we can determine what types of cells or what class

of antibodies are effective against them at these times, we should be able
to develop rational methods of attack.

Antigens

Several facets of the modern biologic revolution have an immense potential for identifying and producing parasite antigens to be used for diagnosis and for vaccines. These include the ability to produce monoclonal
antibodies, the techniques of recombinant DNA which should allow mass
production of some of these antigens, and the direct synthesis of the relevant polypeptide antigens. The use of recombinant DNA and the synthesis of antigens is dealt with elsewhere in this volume, so I will concentrate
on monoclonal antibodies.

Monoclonal Antibodies

Using lymphocyte hybridoma technology, monoclonal antibodies can be
developed which have exquisite specificity. These can be used to identify
parasites directly or to identify and isolate parasite antigens to be used in
serologic assays. Some monoclonal antibodies can passively protect an animal against a challenge infection; these antigens are being used to select

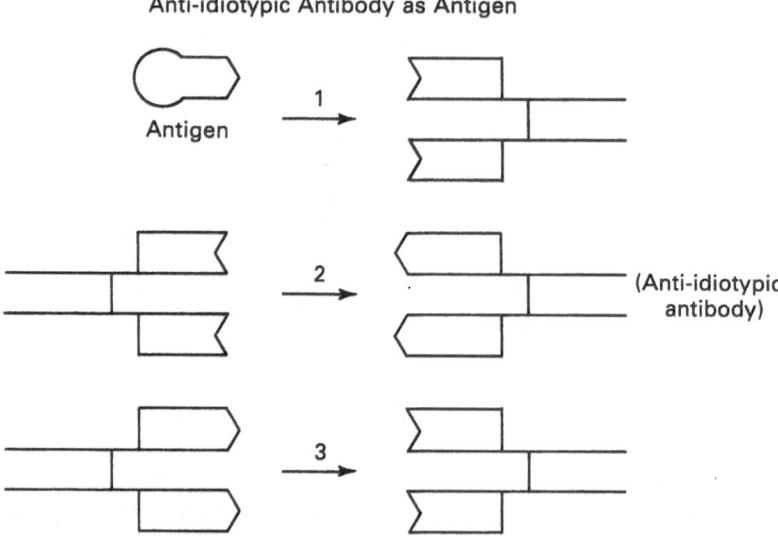

FIGURE 21-1. Anti-idiotype antibody as antigen. In reaction 1, antigen is injected
into a mouse and the spleen is used to produce monoclonal antibodies to that
antigen. In reaction 2, one of the monoclonal antibodies to the antigen is injected
into another mouse and anti-idiotypic antibodies are induced (monoclonal anti-
idiotypic antibodies can also be made). In reaction 3, the anti-idiotypic antibody
is injected into a mouse and induces an antibody which will react with the original
antigen.

putative protective antigens as candidates for vaccine production. It may be possible to detect parasite antigens in the sera of patients with these antibodies, thus allowing the determining of the parasite burden, a parameter which so far is hard to obtain in many parasitic diseases. Drugs or toxins may be conjugated to monoclonal antibodies directed against parasites in order to develop a modern magic bullet.

Finally, monoclonal antibodies may be used to raise anti-idiotypic antibodies, that is, antibodies against the specific antigen combining site of an antibody. It has been postulated that such anti-idiotypic antibodies might be used as "pseudo antigens" to elicit an effective antibody response, bypassing the need for purified antigens themselves (see Fig. 21-1).

Immunopathology

The tissue destruction caused by a number of parasites is thought to be mediated by the immune response of the host. Examples include the immunopathology found in schistosomiasis, filariasis, trypanosomiasis, and leishmaniasis, to name but a few diseases. By understanding the immune mechanisms involved, we may be able to develop specific strategies to diminish the consequences of parasitism.

Although the immunologic approach to the control of parasites is, of course, not the only approach, nor is it necessarily the best one for all circumstances, it is a very important approach which promises to yield significant dividends. We should pursue it with vigor and enthusiasm.

Reference

1. Gross PR (1981) The interface of modern scientific research and parasitology. In: Warren KS, Purcell EF (eds) The current status and future of parasitology. Josiah Macy, Jr. Foundation, New York

22

The Consequences of New Approaches in Parasitological Research

Ruth Arnon and Michael Sela

The title originally assigned to this paper was "The Consequences of the New Parasitology." When contemplating the subject it occurred to us, however, that in reality there is no "new" parasitology. The "old" parasitic diseases are still the same. Regrettably, they have not been eradicated or controlled and still exist with extremely high prevalence. Consequently, the goals of parasitological research also remain the same: development of better diagnostic methods and more efficient means for prophylaxis and cure. Fortunately, new approaches are presently available, and recent advances in molecular biology are being applied for tackling these problems, including modern technologies such as recombinant DNA, cellular hybridization and sophisticated biochemical methods.

It should be stressed that all the experimentation in these directions will be an exercise in futility if it is not accompanied by first-rate epidemiological research and study of vector control. Such modes of research should be combined with the results of laboratory investigation into one concerted effort. The socioeconomic aspects are also very important in the assessment of the tremendous damage created by parasitic diseases. Without mentioning any figures—since this is completely out of our field—it is very clear that the harm to the economy caused by debilitation of the population in endemic countries is one of the reasons for their poverty. Tragically, however, the very same control measures taken to aid development in some of the impoverished nations have had a reverse effect, and irrigation schemes have actually exacerbated several parasitic diseases including malaria, filariasis, and schistosomiasis. What, then, are the expected consequences of employing new approaches to the mainstream of parasitological research?

The six preceding chapters were devoted to a discussion of the future of parasitology, as related to the various fields of research it encompasses, including biochemistry, molecular biology, cell biology, immunology,

246

and epidemiology. The approaches described here are within the same scope, but reflect a somewhat different viewpoint, colored by our own research experience, and providing examples taken from our own work as well as reference to research conducted in other laboratories.

New Approaches to Chemotherapy

The principle of chemotherapy is based upon the use of materials that will be detrimental to the disease-causing agent without having any adverse effect on the host. In the case of the antibacterial field, this has been successfully achieved with the employment of antibiotics. In the field of parasites, this is a much more difficult goal since most antiparasite drugs are of very high general toxicity. Furthermore, in many cases the parasite becomes drug-resistant much before the host. There are three new approaches that can be of help in this situation:

1. The use of antimetabolic drugs which are specific to metabolic pathways that are exclusive to the parasite.
2. Modification of the drug by macromolecularization, in an effort to reduce its toxicity and/or enhance its efficacy.
3. Targeting of the drugs by attachment to carriers with affinity to the parasite.

All three approaches have been attempted, as exemplified in the following:

The first approach was very elegantly demonstrated in Chapter 16 by Dr. Opperdoes, who used the unique glycolytic pathway of trypanosomes as the basis for development of drugs directed at glycolysis as a target.

Another example is the recent development of the new Chinese antimalarial drug Quinghaosu. This drug was isolated and its structure defined, following which several of its derivatives with even higher activity have been prepared. They belong to a completely new series of antimalarial compounds, with a very high level of blood schizonticidal activity against chloroquine-resistant malaria parasites as well, while their adverse effects, manifested only at very high doses, are minimal and reversible. These results illustrate the importance of the investigation of the metabolic processes effective in parasites and their application for the development of new chemotherapeutic drugs.

Macromolecularization of Drugs

This process is another means which may prove of significant pharmacological value, as illustrated by two examples: (1) A newly developed synethetic material, N-acetylmuramyl-L-alanyl-D-isoglutamine (abbreviated as MDP), which contains the minimal active structure in mycobacterial cell wall, was demonstrated to be a very efficient synthetic adjuvant. It also proved active in providing nonspecific resistance to infection. Thus, MDP protected mice from death due to infection with Klebsiella pneumoniae.[1] Macromolecularization of MDP by attachment of several units to a synthetic branched poly-DL-alanine potentiated strongly its capacity to protect the mice from death from infection, but increased significantly also its known pyrogenicity. Surprisingly, an inactive analogue of MDP, in which L-alanine was replaced by D-alanine, becomes—after conjugation to branched poly-DL-alanine under the same conditions—capable of increasing nonspecific immunity to infection although it lacks pyrogenicity.[2] It thus seems that by macromolecularization not only can the pharmacological activity of a drug be retained but it can also be changed for improvement of characteristics.

(2) The anticancer agent daunomycin was also employed in investigation of the effect of macromolecularization. A daunomycin-dextran conjugate (dau-dex) was compared to free daunomycin for acute and subacute toxicity and efficacy in tumor chemotherapy. LD50 and LD2 values of dau-dex were found to be threefold higher than those of the free drug. The therapeutic index of dau-dex was also higher, manifesting a "safe-region" in which no mortality occurs either from the treated lymphoma or from drug toxicity. The subacute toxicity of dau-dex was altogether much lower than that of free daunomycin, as expressed by negligible histologic damage to all organs examined and compared to the massive atrophy of spleen and bone marrow effected by the free drug. Furthermore, in contrast to the known high cardiotoxicity of daunomycin, dau-dex caused almost no damage to the heart tissue. Such a macromolecular conjugate can, therefore, be of benefit for chemotherapeutic use.

Targeting Drugs

Finally the targeting of drugs by the use of carriers that would recognize the parasite *in vivo,* so as to selectively increase the local concentration of the drug while its systemic concentration is maintained at a low, nontoxic, level, may prove highly valuable. Until now this approach has not been employed in research on parasites, but in studies on cancer site-directed chemotherapy it has yielded promising results. The carriers used were varied—ranging from nonspecific lysosomotropic agents such as DNA through liposomes or magnetic microspheres, to specific antibodies. The investigations included several cytotoxic drugs, both small molecules and high molecular weight toxins. Our own experience, with daunomycin linked to antitumor antibodies, indicated that antibodies may serve as carriers for anticancer drugs. The resultant conjugates retain both the pharmacological activity of the drug and the capacity of the antibody to recognize its target and react with it in a specific manner. The conjugates, though of macromolecular nature, are capable of penetrating the target cell and reaching its nucleus. Furthermore, they are capable of "homing" *in vivo* to the tumor site. Consequently, promising results of prolongation of survival of mice as well as complete prevention of tumor growth have been observed in several murine tumor systems as well as in the case of rat hepatoma, using daunomycin conjugates of the respective immuno-specifically purified antibodies.[4] With the rapid development of hybridoma technology and the availability of several monoclonal antibodies reactive with tumor cells, or with relevant parasite antigens, it is to be hoped that the attachment of drugs to such antibodies will result in efficient conjugates for selective destruction of cancer tissue as well as for targeting of drugs to the relevant parasites *in vivo.*

New Approaches to Diagnosis

Diagnosis is one of the most important aspects of the studies of parasitic diseases, not merely for distinguishing the inflicted individuals from healthy ones, but also in epidemiological studies, where a reliable diagnostic method is necessary for evaluation of the results of any chemotherapy field work. For most parasites, several diagnostic tests and techniques are available. However, only very few of them are sufficiently specific and sensitive for detection of low levels of parasitemia, or for distinguishing between an active and nonactive form of the disease. Due to their high specificity and sensitivity, immunodiagnostic procedures are becoming more common than others.

More recently efforts have been devoted to the identification of specific antigens or antigenic determinants of parasites, and they are already being utilized for the development of better immunodiagnostic assays. A few examples that could be mentioned are: (1) The use of purified soluble egg antigen (MSA_1) from schistosome eggs for the development of a reliable, rapid, and simple radioimmunoassay for diagnosis of human schistosomiasis.[5] The antigen, which was later shown to be a glycoprotein, has striking immunochemical and species specificity and appears to be the principal antigen responsible for the granulomatous response to the *S. mansoni* eggs. The reliability of the tests is indicated by the finding that all control sera tested were negative, and 100% of infected samples were seropositive, with very high sensitivity. (2) The recent isolation of two major surface antigens on *Trypanosoma cruzi,* the causative agent of Chagas' disease. One of these, a 75,000 dalton glycoprotein, is specific for the culture forms (insect-host stage) of the organism—epimastigotes and metacyclic trypomastigotes. The other, of 90,000 dalton, was found in vertebrate-host stages of the organisms—bloodstream form trypomastigotes. These two surface antigens are unrelated glycoproteins, as judged by tryptic and chymotryptic peptide analysis. Antibodies were raised in rabbits against these epimastigote and trypomastigote proteins and they were found to react only with the respective homologous immunogens. Therefore, it appears that the insect and vertebrate stages of this species display distinctive surface glycoproteins that could be identified by surface labeling, in six strains of *T. cruzi*, isolated from widely different areas of South America.[6] This finding is of great importance, since such surface antigens can lead to specific antibodies, polyclonal or monoclonal, which will serve for detection of various stages of the parasite *in vivo*.

A new approach in this field is the use of monoclonal antibodies for identification of antigenic determinants. This has been applied with many molecularly defined antigens as well as with viruses and bacteria. A very elegant example for such studies is the investigation of the antigenic drift in Type A influenza virus using variants that had been selected with monoclonal antibodies. In this case the antibodies not only detected the drift but actually induced the formation of new laboratory viral strains.[7] Similar type of results are now expected in studies on parasites.

A case of point is a significant recent work related to a defined antigenic determinant in *T. cruzi*.[8] Here monoclonal antibodies (of the IgM λ class), which were raised against membranes from rat dorsal root ganglia, were found to label *T. cruzi*. These antibodies define a novel antigenic determinant expressed by subpopulation of mammalian central and peripheral neurons, and in the presence of complement they are cytotoxic to mammalian neurons. It is of interest that classes of mammalian neurons and cardiac muscle that are labeled by the monoclonal antibodies are known to degenerate in Chagas' disease. The common neuronal and trypanosomal antigens recognized by these antibodies may therefore be important in the pathogenic events underlying Chagas' disease. On the other hand, they point to the possible danger of tissue damage that can be brought about by vaccination with such cross-reactive antigens, or by the use of such monoclonal antibodies *in vivo*.

Another example is in the field of malaria and consists of the use of hybridoma that produces protective antibodies directed against the sporozoite stage of the parasite.[9] These antibodies, produced by a hybrid of plasmacytoma cells with immune murine spleen cells, were capable of abolishing the infectivity of sporozoites. Moreover, using these antibodies a surface membrane antigen of the sporozoites has been identified. This antigen has protective properties and by metabolic labeling was shown to be identical to a protective antigen previously identified by surface labeling. Characterized as a polypeptide antigen with apparent molecular weight of 44,000, it seems to be strictly associated with only one developmental stage of the parasite. It was not detected in blood forms and was found only in minute amounts in sporozoites from the mid-gut of mosquitoes.[10] This is, therefore, an ideal antigen to be put to use in diagnosis. In consequence these monoclonal antibodies were used recently for the production of anti-idiotypic antibodies, which can in turn be utilized for a very specific and sensitive immunological assay to detect the malaria parasite.[11] Since in this capacity the anti-idiotype antibody actually mimics the antigen, this approach may be developed in the future into very worthy diagnostic procedures, for many other parasites as well. It is especially attractive in the field of parasites, where the preparation of pure antigenic material for the assay presents extreme difficulties.

New Approaches to Vaccination

Synthetic Determinants

The progress in new approaches to vaccination has become in recent years a very tangible and multipronged one. It includes the use of synthetic antigens and synthetic adjuvants, viral subunits and liposomes, as well as antigens prepared by genetic recombination. We will concentrate here on the synthetic approach, and discuss the possibility that vaccines used today might be replaced in the future with antigens containing unique synthetic determinants of disease-inducing agents.

As a result of considerable effort in the last two decades, it has been possible to identify antigenic determinants of many proteins. In most cases these determinants consisted of a relatively small number of amino acid residues with known sequences. With the presently available methodology for peptide synthesis, such peptide fragments can be readily synthesized and used for the preparation of appropriate immunogenic conjugates. This approach has been employed in the case of several protein antigens and has led to the production of synthetic antigens capable of eliciting specific antiprotein immune responses.[12]

In our own laboratory we have shown that antibodies provoked in this manner by a conjugate containing the synthetic peptide corresponding to the "loop" region of the enzyme lysozyme were directed to a conformation-dependent determinant, and they reacted with the intact native protein.[13] In view of this we anticipated that it should be feasible to employ a similar approach for components of viruses or bacteria, for induction of neutralizing antibodies. Indeed, we prepared an immunologically active synthetic peptide that is analogous to a fragment from the coliphage MS-2. A conjugate of this peptide with a synthetic carrier, administered in Freund's adjuvant, provoked in rabbits and guinea pigs antibodies capable of efficiently inactivating the native bacteriophage.[14] Furthermore, when the synthetic adjuvant N-acetylmuramyl dipeptide (MDP) was attached to the above conjugate, it yielded a completely synthetic molecule that induced high anti-MS-2 neutralizing response even when injected in phosphate-buffered saline.[15]

A recent publication by Audibert and colleagues[16] reported that active antitoxic immunization against diphtheria could also be achieved with a synthetic peptide covalently linked to a protein carrier. This is a tetradecapeptide and represents a fragment of the loop (14 amino acids sustained by two cystenics) which delineates the two functional fragments of the natural diphtheria toxin molecule. Injection of the conjugate into guinea pigs elicited protective activity which neutralized the dermonecrotic action of the toxin.

The chemical approach was proven effective also for some animal viruses such as hepatitis and influenza. In the case of hepatitis B, peptides with sequences covering most of the virus envelope protein molecule, particularly the hydrophilic domains, have been synthesized.[17,18] Several of these peptides induced in rabbits an immune response toward the native envelope protein of Dane particles of the hepatitis B virus. In the case of influenza, we have synthesized a fragment of the hemagglutinin molecule of influenza, which comprises a common sequence for several strains of influenza subtype A. In rabbits, a conjugate of this peptide with tetanus toxoid induced antibodies that reacted immunochemically with the peptide as well as with the intact virus of the relevant strains. The antibodies inhibited the hemagglutinin activity of this virus, and also interfered with its plaque formation *in vitro.* Moreover, immunization of mice with this conjugate resulted in partial protection against a challenge infection.[19]

It is hoped that further research directed toward expansion and optimalization of this approach will lead to the safest and most effective vaccines of the future toward various disease-causing agents, including parasites. This will be a crucial step forward, especially in the case of parasites, where natural material to be used for vaccination is so hard to come by.

Monoclonal Protective Antibodies

Another approach to vaccination against parasites is one mentioned earlier, namely, the use of monoclonal protective antibodies for identification and possible isolation of relevant antigens that might lead to protective immunity. An alternative approach would be the use of anti-idotype antibodies to such monoclonal antibodies, which might mimic the antigen and thus replace it for vaccination purposes. First attempts in this direction have already been made by Graham Mitchell and by Alan Sher. This idea, similar to the synthetic approach, is extremely attractive for application in the field of parasitic diseases, in view of the difficulties involved in the preparation of antigenic material for vaccine production.

"Specialized" Immunization

The last notion to be described here is the possible employment of specialized immunization that might lead to protective immunity, and in this context an example may be mentioned from our own work on schistosomiasis. This parasite induces in the host a strong humoral and cellular immune responses, but no protective immunity. The reason for that might be the known resistance to immune attack the schistosomula develop spontaneously within short time after conversion. This led us to the hypothesis that protective immunization might be achieved if effective during the first hours after penetration, namely while the parasite is still in the skin. Since skin-bound antibodies belong mainly to the IgE class, we tested the hypothesis by using immunization procedures known to elicit specific IgE antibodies. Indeed, a partial protection of mice was achieved by immunization with a minute amount of cercarial antigen in alum. A decrease of 34–90% in the adult worm burden of the immunized mice as compared to control groups was the result, and the site of attrition was demonstrated to be in the skin. The protection was paralleled by high levels of specific anticercarial antibodies of the IgE class. It is thus possible that specific IgE antibodies play an important role in protective immunity against schistosomiasis. In the case of other parasitic diseases, parallel approaches could be envisaged for such "specialized" immunity.

Progress in all the approaches just described will give us new formidable challenges, because ultimately the new concepts can help us move ahead. But only the perseverance to translate these concepts into reality will bring us closer to the expected goals, and to a genuinely "new parasitology."

References

1. Chedid L, Parant M, Parant F, Lefrancier P, Choay J, Lederer E (1977) Enhancement of nonspecific immunity to *Klebsiella pneumonia* infection by a synthetic immunoadjuvant (N-acetylmuramyl-L-alanyl-D-isoglutamine) and several analogs. Proc Natl Acad Sci 74: 2089.
2. Chedid L, Parant M, Parant F, Audibert F, Lefrancier P, Choay J, Sela M (1979) Enhancement of certain biological activities of muramyl dipeptide derivatives after conjugation to multi-poly (DL-alanine)–poly (L-lysine) carrier. Proc Natl Acad Sci 76: 6558.
3. Levi-Schaffer F, Bernstein A, Meshorer A, Arnon R (1982) Reduced toxicity of daunomycin by conjugation to dextran. Cancer Treatment Reports 66: 107.
4. Arnon R, and Sela M (1982) In vitro and in vivo efficacy of conjugates of daunomycin with anti-tumor antibodies. Immunol Rev 62: 5.
5. Pelley RP, Warren KS and Jordan P (1977) Purified antigen radioimmunoassay in serological diagnosis of *Schistosomiasis mansoni*. Lancet 2: 781.
6. Nogueira N, Unkeless J, Cohn Z (1982) Specific glycoprotein antigens on the surface of insect and mammalian stages of *Trypanosoma cruzi*. Proc Natl Acad Sci 79: 1259.
7. Laver WG, Gerhard W, Webster RG, Frankel ME, Air GM (1979) Antigenic drift in type A Influenza virus: Peptide mapping and antigenic analysis of A/PR/8/34 (HON1) variants selected with monoclonal antibodies. Proc Natl Acad Sci 76: 1425.
8. Wood JN, Hudson L, Jessel TM, Yamamoto M (1982) A monoclonal antibody defining antigenic determinants on subpopulations of mammalian neurones and *Trypanosoma cruzi* parasites. Nature 296: 34.
9. Yoshida N, Nussenzweig RS, Potocnjak P, Nussenzweig V, Aikawa M (1980) Hybridoma produces protective antibodies directed against the sporozoite stage of Malaria parasite. Science 207: 71.

10. Yoshida N, Potocnjak F, Nussenzweig V, Nussenzweig R (1981) Biosyntheses of Pb$_{44}$, the protective antigen of sporozoites of *Plasodium berghei*. J Exp Med 154: 1225.

11. Potocnjak P, Zavala F, Nussenzweig R, Nussenzweig V (1982) Inhibition of idiotype-anti-idiotype interaction for detection of parasite antigen; A new immunoassay. Science 215: 1637.

12. Arnon R (1980) Chemically defined anti-viral vaccines. Ann Rev Microbiol 34: 593.

13. Arnon R, Maron R, Sela M, Anfinsen CB (1971) Antibodies reactive with native lysozyme elicited by a completely synthetic antigen. Proc Natl Acad Sci 68: 1450–1455.

14. Langbeheim H, Arnon R, Sela M (1976) Anti-viral effect on MS-2 coliphage obtained with a synthetic antigen. Proc Natl Acad Sci 73: 4636–4640.

15. Arnon R, Sela M, Parant M, Chedid L (1980) Anti-viral response elicited by a completley synthetic antigen with built-in adjuvanticity. Proc Natl Acad Sci 77: 6769–6772.

16. Audibert F, Jolivel M, Chedid L, Alouf JE, Boquet P, Rivaille P, Siffert O (1981) Active antitoxic immunization by a diphtheria toxin synthetic oligopeptide. Nature 289: 593.

17. Lerner RA, Green N, Alexander H, Liu F-T, Sutcliffe JG, Shinnick TM (1981) Chemically synthetsized peptides predicted from the nucleotide sequence of the Hepatitis B virus genome elicit antibodies reactive with the native envelope protein of Dane particles. Proc Natl Acad Sci 78: 3403–3407.

18. Dreesman GR, Sanchez Y, Ionescu-Matin I, Sparrow JT, Siv HR, Peterson DL, Hollenger FB, Melnik JL (1982) Antibodies to Hepatitis B surface antigen after a single inoculation of uncoupled synthetic HBsAg peptides. Nature 295: 158.

19. Müller G, Shapira M, Arnon R (1982) Anti-influenza response achieved by immunization with a synthetic antigen. Proc Natl Acad Sci 79: 569.

20. Horowitz S, Smolarsky M, Arnon R (1982) Protection against *Schistosoma mansoni* achieved by immunization with sonicated parasite. Eur J Immunol 12: 327.

Discussion

SELL: Dr. Arnon, the carrier you used was tetran, and the adjuvant was completely Freund's adjuvant, so isn't it true that there was no NDP and the part you prepared was specifically for the H-3 entry, not for the other entries?

ARNON: That is right.

SELL: The reason I asked these questions specifically is that the new model points out the potential of this method for use in parasitology. It also points out the difficulty, however, because the new virus can change itself periodically whenever we think we have it licked. We can synthesize whatever you like, but if the new virus changes its hemagglutinin, it can go on to affect the next generation.

One must synthesize something that is common to hemagglutinins of all H-types if these common areas prove to be immunogenic, and we hope they will be.

Earlier in the conference, a molecular biologist said the idea of immunizing with trypanosomes seemed somewhat remote because of the changing antigen on genetic analysis. Yet if one synthesizes a portion of the antigen that is common and consistent, even though it ordinarily does not produce immunity or an antibody, it is possible to produce immunity even with an organism that is trying to change its antigenic structure to hide itself.

So one could pick a portion of the antigen, synthesize it, and produce an antibody never produced before, and yet have it give protection.

These are both the challenges, the opportunities, and the potential pitfalls of using the system, but I think the artificial antigen system is very exciting.

Mitchell: Another issue that should be raised is the probability that one should start immunizing with very defined antigens, and the probability of selecting for antigen-negative parasites. I have never known of a parasite infection where one would vaccinate with anything resembling these kinds of small peptides to induce immune responses against this complex organism.

I can envisage in many situations, exactly the same as in the trypanosomes, that if you have in malaria a very small peptide, which is the portion of an immunogen required for the immune response that gives protection, there will automatically be selection. Therefore, in the new biology and in the new vaccination approach we will be much in the same boat as in chemotherapy, where there is very rapid resistance to our current strategy.

Soprunov: Was it strain specific?

Arnon: It was specific to the subtype of the virus that contained it; we did not do it *in vivo*.

International Networks for Research in Parasitology

23

The WHO Network

Adetokunbo O. Lucas

The Special Programme for Research and Training in Tropical Diseases was initiated by the World Health Organization following a resolution of the World Health Assembly calling for the intensification of research into tropical diseases, and is co-sponsored by the United Nations Development Programme and the World Bank. The Programme has two interrelated objectives:

Research and development for better tools to control tropical diseases, and
Strengthening of institutions, including those providing training, to increase the research capability of tropical countries.

The mechanisms evolved by the Programme place considerable emphasis on the establishment of collaborative activities in national institutions which have been organized into functional networks. In this paper, I will present a brief summary of the operation of these networks.

First it is important to point out that the ultimate goal of the Programme is the improved control and the eventual conquest of tropical diseases, with emphasis on the six diseases which have been selected for initial attack: malaria, schistosomiasis, filariasis (including onchocerciasis, or river blindness), trypanosomiases (both African sleeping sickness and the American form called Chagas' disease), leishmaniases, and leprosy.

It is also important to note that the programme has been designed to work closely with governments, universities, research institutes, and the pharmaceutical and chemical industries. Thus, the Programme does not attempt to and could not work in isolation. Rather it must collaborate with national, bilateral, and international efforts and, whenever appropriate, should serve a coordinating role.

It should also be noted that the Programme does not represent WHO's first venture into research in this field. Since its inception, WHO has actively promoted and conducted research in malaria, schistosomiasis, and various other priority health problems of the tropics. The Special Programme has been able to draw on this accumulated experience.

Research and Development

The research and development component of the Programme is planned, implemented, and evaluated by multidisciplinary teams of scientists who are selected on the basis of their expertise and experience. So far, a total of 2,300 scientists from 118 countries have been involved in Special Programme activities. How are these networks operating? The question is best answered by examining a few concrete examples.

Filariasis

The Scientific Working Group (SWG) on Filariasis, which was established in 1977, gave priority to the problem of onchocerciasis, but is now expanding its work on lymphatic filariasis. Immunological research is aimed at identifying filarial antigens to be used in serodiagnostic tests, and perhaps for the development of vaccines. The Group is also supporting work on the natural history, epidemiology, and vectors of filarial infections with a view towards improving methods for controlling their transmission. With regard to chemotherapy and drug development, the Group aims to improve the use of existing filaricides and to find new ones; it seeks also to find means of reducing the inflammatory reactions that occur in the human host in response to the presence and death of filarial worms.

In the search for new drugs, the scope of activities in the Programme ranges from the study of metabolic pathways in filarial worms, to the synthesis of chemical compounds around existing and new leads, the identification of substances with filaricidal activity in various biological screens, the clinical trials of promising compounds and field trials.

METABOLIC STUDIES. So far the metabolic studies have shown interesting features of folate metabolism in filarial worms—the ability of adult worms to oxidize 5-methyl tetrahydrofolate directly to 5,10-methylene tetrahydrofolate—and their dependence on *de novo* synthesis of thymidilate and on exogenous methionine. The effects of suramin on the enzymes of the glycolytic pathways in *Onchocerca volvulus* are also being investigated, with early results indicating that the therapeutic effect of the drug includes the blockage of reduction and reoxidation of pyrimidine nucleotide, thereby inhibiting energy metabolism. These metabolic studies are being pursued in the hope of identifying pathways peculiar to the worms which would be suitable targets for specific drug action.

SYNTHESIS. The SWG on Filariasis commissioned a study on the structure–activity relationship of filaricidal compounds. The findings of this study and the results of metabolic studies have guided the Programme for the synthesis of new antifilarial compounds.

SCREENING. A network of biological screens is being supported in several countries—the United States, the United Kingdom, Germany, Japan, and Australia. Chemical compounds specifically synthesized for the Programme and other submitted by various pharmaceutical companies are put through the primary and secondary screens, and the most active compounds go to the tertiary screen of cattle onchocerciasis *(Onchocerca gibsoni, O. guttorosa)*. A tertiary scrren using leaf monkey is also being developed for lymphatic filariasis. So far, over 7,000 compounds have been put through the various screens and a number of promising leads have been identified. For example, a combination of mebendazole and 1-tetramisole suppressed embryogenesis in the cattle screen. This combination is being evaluated in man. Several other leads have emerged including compounds which are macrofilaricidal.

CLINICAL WORK. Chemotherapeutic trials are being carried out on existing and new filaricidal drugs at several centres in the endemic areaas—Sudan, Nigeria, Ghana, and Togo. More detailed pharmacological studies are being conducted in collaboration with institutions in developed countries. Such collaboration between the clinician at the hospital in Tamale, Ghana, and scientists in Liverpool, England, has yielded new information about the two drugs, suramin and di-ethylcarbamazine, which are in common use in the treatment of onchocerciasis. Table 23-1 is a brief summary of the impact of this network on the search for a new drug against onchocerciasis.

TABLE 23-1. New drug for onchocerciasis: status report

Criteria	1977	1981
Animal screens available	Minimal	Cattle screen validated Enlarged global screening network in operation
New leads under investigation	None	At least five new chemical groups
Research and Development activity in industry	Small scale and declining	*Ten* pharmaceutical companies* with active R&D programmes

*Located in France, Federal Republic of Germany, Japan, Switzerland, United Kingdom, and the United States.

This example illustrates several features of the Special Programme networks:

1. *Goal orientation.* A specific definable goal has been identified, i.e., the search for a drug which will kill adult onchocercal worms in man.
2. *Strategic planning.* From a thorough review of the literature and of ongoing activities, a detailed plan of work was prepared, including several interrelated components:
   a. comparative biochemistry of filarial worms,
   b. synthesis of chemical compounds,
   c. biological screening of candidate agents,
   d. clinical evaluation, and
   e. field trials.
3. *Golbal collaboration.* This involves academia, industry, and governments. The fact that WHO is a global organization with 158 member states facilitated this collaboration. For research and development, the strategic plans are being implemented by scientists working mainly in their own institutions, their activities being coordinated into functional networks.

This network approach is proving productive and cost-effective. It has provided the means for overcoming constraints which in the past have limited research work on some of these diseases. By providing scarce biological materials, a number of scientists have been able to contribute their specialized skills—in disciplines such as immunology, molecular biology, biochemistry—to the research effort. For example, through the Immunology of Leprosy Scientific Working Group (IMMLEP), the nine-banded armadillo from Central America is providing scientists throughout the world with leprosy bacilli for experimental studies. Similarly, other Working Groups have offered and provided freeze-dried worms (schistosomes) and sera from patients and controls (Chagas' disease). Through the network mechanism, it has been possible to tackle problems affecting many countries, using standard protocols (e.g., epidemiology of leishmaniasis) and standard diagnostic kits (e.g., *in vitro* testing of the sensitivity of malarial parasites to drugs). The network concept has made it possible to conduct experiments and trials where the local ecological situation is most ideal for answering specific questions. It has been possible to exploit in Australia the natural onchocercal infection in cattle as the tertiary and definite model for the human disease which occurs only in tropical Africa and Central America. The enthusiastic collaboration of scientists throughout the world, with the generous cooperation of their institutions and the willing agreement of their governments, has made it possible to exploit these various opportunities on a global scale.

Similar progress has been made in other areas of the Programme.

Malaria

A total of 88 projects has been funded on chemotherapy and drug development (CHEMAL), 69 on immunology (IMMAL), and 61 on field research (FIELDMAL).

Under CHEMAL, work has continued on the development of mefloquine with clinical trials in Brazil, Thailand, and Zambia. In association with Chinese scientists, progress has been made with the preclinical development of Qing Hao-su (artemisinine) and its derivatives, some of which have shown significantly higher activity than the parent compound. Work has continued on the development of kits for performing the micro-test for the sensitivity of malaria parasites to drugs. Such expermental kits are currently being evaluated in endemic countries.

The development of tissue schizonticidal drugs was reviewed at a recent meeting and priorities for research were identified. Work continued on the metabolic pathways in malaria parasites, with some comparisons among strains which differ in their response to drugs.

The IMMAL programme has received a major stimulus through the introduction of the cell-fusion (hybridoma) technique for the production of monoclonal antibodies. Several institutes, both inside and outside the programme, have taken up this line of research, with particular emphasis on parasite-inactivating or growth-inhibiting antibodies, especially those interacting with surface antigens of sporozoites, merozoites, and gametes. A protective monoclonal antibody directed against a sporozoite surface antigen of *P. berghei* has been produced and used for the isolation of the relevant antigen which is undergoing further analysis, with a view towards eventually reproducing and mass producing the antigen through modern methods of genetic engineering. At the same time, the system is being applied to *P. falciparum*. As a result of the recent advances in hybridoma and recombinant DNA techniques, the concept of vaccine development has shifted from the use of crude whole-parasite to that of defined protective antigens. Besides the investigation on the mechanisms of immunity, immune evasion, immunosuppression, and immunopathological phenomena, IMMAL has supported the further development of immunodiagnostic tests. A solid-phase radioimmunoassay for the detection of low numbers of malaria parasites in blood has been adapted from a rodent model to *P. falciparum*. It detects parasites down to a level of 8 parasites per 10^6 RBC. Current work is directed at the adaptation of the test to an ELISA system.

Research in the area of parasite biology is being pursued jointly by CHEMAL and IMMAL since both require the support of this discipline. Further progress has been made in the *in vitro* cultivation of blood stages of *P. falciparum* and other plasmodia which should eventually lead to the culture of *P. malariae* and *P. vivax*. The *in vitro* cultivation of exoerythrocytic stages of *P. berghei* from sporozoites to infective merozoites, achieved outside the programme, will facilitate the development of *in vitro* screening systems for tissue schizonticidal compounds and, for the first time, render feasible studies on sporozoite invasion. The production of viable gametocytes *(P. falciparum)* from *in vitro* culture has opened a way to raising sporozoites in the numbers needed for chemotherapeutic and immunological research. Besides these studies, particular emphasis was given to membrane studies—of both erythrocytes and parasites—to elucidate structure and function in relation to invasion, material and energy transport, and antigenic compounds.

In response to the most urgent problem facing malaria control in widespread parts of East Asia and South America, FIELDMAL has given high priority to global studies on the drug resistance of malaria parasites, especially *P. falciparum*. These studies are carried out in close cooperation with the malaria services of the countries concerned, and the WHO Regional Offices and the SWG/SC Secretariat. Besides the assessment of baselines and the monitoring of drug sensitivity levels, major efforts are directed at the development and strengthening of methods for the containment of drug-resistant malaria. These activities have been consolidated in the countries of the Southeast Asia and Western Pacific Regions and some progress has been made in implementing them in other regions. Further essential research activities of FIELDMAL are focused on studies of the operational use of antimalarial drugs and community participation in antimalarial measures, vector control in areas with exophilic or insecticide-resistant anophelines, and epidemiological research as the basis for rational planning and evaluation of malaria control. Special efforts are being made in research training and in the improvement of field research capabilities of national antimalarial services and scientific institutions in tropical, malarious countries.

Schistosomiasis

Three major areas have been included in the programme:

1. Applied field research, including vector biology and control,
2. Chemotherapy, and
3. Immunology.

So far, this component has funded 110 projects. With regard to applied field research, work has been done on integrated approaches to the control of schistosomiasis in man-made situations—large dam lakes and irrigation systems. The findings from the studies at Lake Volta will be published soon. The use of focal mollusciciding based on ecological requirements is being investigated in the Sudan, as is the study of the dynamics of snail transmission in different irrigation systems. In another study, the impact of chemotherapy on high-density *S. mansoni* infection is being examined. Eleven physicians participated in a workshop on population epidemiology of *S. japonica* in the Philippines.

The SWG has also sponsored research on the intermediate snail hosts of schistosomiasis, including studies on ecology, snail behaviour and the role of attractants, the genetics of resistance to infection with schistosomes, and the life cycle of the parasite within the snail. Metabolic pathways in snails are being explored in the hope of identifying suitable targets for molluscicidal action. A slow-release formulation of a molluscicide is undergoing laboratory development and testing.

Projects have been funded on the chemotherapy of schistosomiasis complementary to the efforts of the pharmaceutical industry. Some basic biochemical studies have been designed to elucidate the mode of action of schistosomicidal drugs as well as the pharmacological properties of these agents in man. Niridazole and metrifonate have been examined in this way. Clinical evaluation of new drugs has also been supported with trials on praziquantel, a drug developed by industry in collaboration with the Schistosomiasis Unit of WHO.

Immunological research is being carried out with full knowledge of the work sponsored by the Edna McConnell Clark Foundation and, where relevant, collaborative work has been organized. A collaborative study for immunodiagnosis involving eight laboratories has recently been completed in collaboration with this Foundation. The SWG has also promoted research in this field of immunology by providing parasite material, for example, lyophilised adult worms and eggs of *S. mansoni*. The immunological responses of the mammalian host to bilharzia infection are being studied, including work on the mechanism of resistance to infection.

African Trypanosomiases

The work of this component is organized in three subsections:

EPIAF—epidemiology and control
CHEMAF—chemotherapy and drug development
IMMAF—immunology

Recent research results have promoted understanding of the epidemiology of this disease. Several game animals have been shown to harbour trypanozoon stocks, some of which appear to be identical to human stock of *Trypanosoma b. gambiense*. This complements the earlier finding of similar infection in domestic animals—pigs and dogs. Further confirmation has been obtained of the earlier finding that tsetse flies travel over much longer distances than had been assumed in the past. Another field trial of the ion-exchange minicolumns for parasitological diagnosis was carried out, this time in a *T.b. rhodesiense* endemic area. The direct card agglutination test for trypanosomiasis (CATT) has been further improved to achieve better fixation of the antigen to the card. In collaboration with the SWG in Epidemiology, a longitudinal study commenced in Zambia.

With regard to drug development, interest has been shown in compounds which can disrupt threonine metabolism and those which affect the enzyme ornithine decarboxylase of the parasite. Screening of potential trypanocides continues at two centres, one in Kenya and the other in the Federal Republic of Germany. Pharmacological studies continue on the drugs in current use—Antrypol and organic arsenicals.

Although immune complexes are found in the sera of patients with and without cerebral complications in the cerebrospinal fluid, complexes were found only in the meningoencephalitic stage. Work continues on the antigenic variation in these parasites, with regard to the repertoire and also the mechanism for its occurrence.

Chagas' Disease
The work of the Chagas' Disease SWG is organized in three major areas:

EPICHA—the epidemiology of Chagas' disease and the studies of the vectors
CHEMCHA—chemotherapy and drug development
IMCHA—immunology of Chagas' disease

An important objective of the SWG is to improve knowledge of the geographical distribution, prevalence, and clinical varieties of Chagas' disease and of the distribution of its vectors. Field research on the prevalence and distribution of the disease has been initiated in several endemic countries. Studies using standard protocols and methods have been initiated in three countries. Some results have been obtained on the analysis of blood meals as a means of assessing the relative role of potential vectors in different geographical areas. A field evaluation of a slow-release polyvinylchloride-based paint showed promising results with effective control of the insect vector for nine months. Work has continued on the study of the mode of action of organophosphorous compounds on *Triatoma infestans.* Work on the metabolic pathways of *Trypanosoma cruzi* continues to yield interesting results including studies on the salvage and interconversion of purines, *de novo* biosynthesis of pyrimidine, and analysis of the respiratory chain. Several potential trypanocides have been identified, and more compounds are being screened.

A comparative serological study was initiated in July 1980 and national laboratories of six Latin American countries are currently collaborating in the project. In a project based in Brazil, reference sera from patients and uninfected controls have been collected and are being made available to scientists for standardization of their tests. Mechanisms of the pathogenesis of Chagas' lesions are being elucidated with particular interest in the role of antibodies against endocardium, vascular structures, and the interstitium of heart and striated muscles, as well as those antibodies active against peripheral nerves. Focal lesions resulting from intracellular parasitism of muscle cells have also been reported.

Work on antigenic analysis is making good progress; specific antigenic components are being identified by the use of monoclonal antibodies.

Leishmaniases

One section of the SWG has been dealing with epidemiology, parasitology, vector biology, and control of leishmaniases. Much emphasis has been placed on increasing knowledge about the geographical distribution and varieties of these diseases. The existing literature has been reviewed and is being summarized on a country-by-country basis. Epidemiological surveys are currently in progress in 17 countries, and the preliminary data obtained will be examined and analysed later this year. The animal reservoirs of both the cutaneous and mucocutaneous visceral forms are being identified in different geographical locations.

Type collections of sandflies from many different parts of the world are being catalogued at the British Museum for taxonomic purposes and for the training of scientists. Blood meal identification is being widely used to incriminate local vectors.

Efforts continue to be made to obtain more exact typing of leishmanial strains. In addition to the use of standard serological, biochemical (enzymological), and biological characteristics, a new technique using radiorespirometry has provided informative results. Although the latter method is too complex for routine use, it could prove valuable for definitive classification of reference material.

Clinical evaluation of promising drugs has continued. It was again confirmed that nifurtimox shows some effect in cases of mucocutaneous leishmaniasis, but on its own the drug did not achieve a high cure rate. In combination, it improved the results obtained by treating with meglumide antimoniate alone. Allopurinol, which had shown *in vitro* activity, has so far proven disappointing in clinical trials in cases of visceral leishmaniasis. The observation in experimental animals that entrapment in liposomes enhanced the therapeutic activity of some antileishmanial drugs was initially made by scientists working outside the programme, and this approach is now receiving support in the hope of applying this technology in human disease. A modest screening effort is continuing and interesting activity has been identified with a few compounds. Meanwhile, biological screens for cutaneous and visceral disease are being developed and evaluated. In a series of experiments on the problem of treatment failures, the strains of *L. donovani* involved exhibited resistance to the drug meglumide antimoniate. At a workshop on the chemotherapy of mucocutaneous leishmaniases, agreement was reached on standardized protocols for drug trials including the selection of patients, identification of the parasites, treatment schedules, follow-up, and criteria of cure.

The immunology section has supported work on diagnosis using modern techniques of antigenic analysis and specific monoclonal antibodies. Mechanisms of immunity are also being studied. With the demonstration of cross-reacting antigens between *L. enriettii* and *L. tropica,* the possibility of using the former nonpathogenic species as a vaccine is being explored.

Leprosy

One SWG, IMMLEP deals with the immunology of leprosy and the other, THELEP, with chemotherapy and drug development.

Work on the development of the antileprosy vaccine continues to make good progress. Experiments in mice confirm the protective effect of four different preparations of killed *Mycobacterium leprae.* The preferred procedure (protocol 1/79) produces a high yield, with minimal damage to bacteria and minimal contamination. It is proposed to undertake the testing of the immunogenic potential of this preparation with and without BCG in human beings as the first step in the vaccine trials.

Immunodiagnostic tests are being further developed. An ELISA test of comparable sensitivity to the radioimmunoassay test to *M. leprae* has been established. A method has also been developed for early detection of systemic infection in armadillos, and monoclonal antibodies are being evaluated for their specificity for *M. leprae.*

Clinical trials of drug combinations continue under the auspices of THELEP. At one centre, several cases of jaundice led to the suspension of one regime but investigations suggest that drug toxicity was not cause of the jaundice. Detailed protocols for field trials of chemotherapy of lepromatous leprosy were drafted and two trials are expected to start soon. With regard to drug development, analogues of thalidomide and of rifampicin failed to yield promising leads, but work continues on analogues of ethionamide and prothionamide. Work continues on the development of prolonged-release preparations of dapsone. Surveys are being carried out on the frequency of primary dapsone resistance in endemic countries. One report from the Philippines showed a prevalence of 2.1% of new cases of lepromatous leprosy. Results from other geographical areas are expected soon.

Biomedical Sciences

The SWG continues to be active in stimulating further application of basic biomedical sciences to the study of tropical parasitic and infectious diseases, with particular emphasis on approaches which are innovative. The SWG has funded projects dealing with genetics and molecular biology. The ongoing projects include the use of recombinant DNA technology for the study of kinetoplast DNA in *Trypanosoma lewisi* and of the variant antigen sequences of Trypanosomatids. Work is progressing on the gene organization and function of parasitic protozoa. The role of factors under genetic control in relation to susceptibility to infection is being investigated with regard to G-6-PD deficiency and *P. falciparum* infection in man. Two studies in animal models are investigating genetic factors in relation to leishmanial infections. Studies are being carried out on the role of cell surfaces and recognition phenomena. The metabolic pathways of parasites are being investigated, including purine metabolism in trypanosomes, the role of oxygen reduction products in the killing of parasites, and the sensitivity of variant forms of superoxide dismutase to cyanide.

The SWG continues to promote the exchange of scientific information through co-sponsorship of courses and workshops.

Biological Control of Vectors

In pursuing its objectives, this component of the programme has collaborated with national scientists as well as with industrial firms in the testing and development of biological agents for vector control. The tests for efficacy and safety of the agents have followed the scheme developed by WHO Expert Committees. Of the microbial agents, the highest priority has been assigned to *Bacillus thuringiensis,* serotype H-14, which has now reached the stage of large-scale testing for the control of mosquito and blackfly larvae.

Tests have shown that it is a potent nonresidual larvicide of mosquito and blackfly larvae, with a large safety margin for man and other nontarget organisms. The activity of the agent is not affected by salinity, pH within reasonable limits, and water temperature, but is less effective in polluted than in clear water. The delta endotoxin is stable under tropical conditions.

Work continues with another bacterial agent, *B. sphaericus.* Safety tests show that this agent is innocuous for mammals under normal conditions of exposure, and environmental studies show no harmful effects on nontarget organisms. The agent is pathogenic for larvae of culex and certain species of *Anopheles,* but much less effective against *Aedes.* For the further development of the agent, improved formulations and reliable standardization methods are required.

Culicinomyces clavosporus, a fungal agent, is also being studied as a potential biological agent for the control of mosquito and certain other insect larvae.

Of the nonmicrobial agents, the SWG has assigned a high priority to the fish *Gambusia affinis.* The use of this and other larvivorous fishes will be reviewed at a special consultation later this year. A variety of other agents are being systematically examined according to the standard scheme for investigating potential biological agents.

Epidemiology

The SWG on Epidemiology is seeking to promote research that will contribute to the design of the most effective strategies for the control of tropical parasitic and infectious diseases. A multidisciplinary longitudinal epidemiological study in Zambia is providing useful information about the health status of the population in rural areas. Analysis of the data is showing some interesting correlations suggesting important interactions of malaria and schistosomiasis infection. In collaboration with the SWG on African trypanosomiases, a field study has been initiated and about a thousand persons were examined during the initial survey. The SWG continues to place emphasis on the training of epidemologists and is collaborating with the RSG on the strengthening of training courses at key institutions in developing countries.

In collaboration with the WHO Regional Office for the West Pacific, a regional workshop was held for teachers of epidemiology in Manila. The 26 participants will be polled in a year's time as a follow-up of the workshop. Others support for training in epidemiology includes the preparation of a field manual and the development of simple computer-assisted simulation exercises.

Social and Economic Research

The SWG on Social and Economic Research has focused on the major objective of increasing the effectiveness of disease control programmes through the integration of human behavioral factors in programme design and management. In this context, behaviour was defined to include social, cultural, and economic factors. Most of the projects funded so far relate to the intermediate objective of defining the relationship between these factors and the transmission of diseases and their control. As far as possible, the projects are being developed in association with ongoing epidemiological research or control programmes. Linkages with the work of other SWGs and the RSG are being expanded.

Early results from funded projects include a preliminary report on knowledge, attitudes, and behaviour with regard to malaria. In this study, the scientists sought the reasons for declining collaboration of the population with indoor spraying of DDT in the control programme in Thailand. In another study, the role of the school in the control of locally endemic diseases was examined through a questionnaire administered to primary school children in Nigeria. Human vector contact in relation to African sleeping sickness is being studied in an endemic area of Upper Volta.

The Strengthening of Research Capability

The objective of the Research Capability Strengthening Area of the Special Programme is to assist developing countries where the six diseases are endemic to assume their appropriate role in the research required to identify, analyse, and solve the health problems caused by these diseases. The goals and activities of this Programme Area are interdependent with those of the Research and Development Area, although there are important differences in the nature of the projects supported.

The scope of activities has increased rapidly during the reporting period; 26 institutions are now receiving support on a long-term basis while another 20 have received short-term and capital grants. In addition, other institutions are being supported on a long-term basis to enable them to conduct formal courses at a Master's degree level. The support of training activities also has expanded and over 240 scientists have received individual research training grants and approximately 40 short-term group learning activities have been supported in endemic countries. Over 130 trainees have now returned to their home institutions in developing countries after training, and about one-quarter of that number are being supported by re-entry grants to enable them to apply their knowledge and skills in local situations.

The strategic plan for this area of the Programme, which was formulated and implemented in 1979 and 1980, was reviewed in the light of progress during the past two years and of the findings of the Scientific and Technical Review Committee which carried out an in-depth review of the programme area. The development of a durable network of institutions in endemic countries, which is an objective of this plan, involves a strategy of shifting resources to less developed institutions, while those which initially received support assume responsibility for their own research and training activities. This has been illustrated by the Ndola Tropical Diseases Research Centre, which initially started as a WHO activity. As of 1 January 1981, however, the Government of Zambia assumed responsibility for the management of the Centre, and a Zambian physician/scientist was appointed as its first director. The Centre continues to collaborate with the Special Programme for research in epidemiology and clinical pharmacology; it is also playing an important role in the training of scientists from other developing countries.

This is but one of a number of examples of the policy of promoting technical cooperation among developing countries. Most of such cooperation has occurred through training activities which can lead to the forging of links among developing-country institutions. Of the 53 scientists supported by research training grants and visiting scientist grants between 1 July 1980 and 30 June 1981, 21 will have undertaken all or most of their training in developing countries other than their own. The host countries are: Brazil, Ethiopia, Ivory Coast, Kenya, Thailand, Malaysia, Singapore, Venezuela, and Zambia.

Interaction between scientists from developing countries also occurs through the short-term workshops/seminars and long-term courses. All these group learning activities were located in developing countries and were planned and implemented by local scientists; all included participants from other developing countries. Seven of the institutions receiving long-term support are actively engaged in group training activities involving nationals from other developing countries.

Activities in the Research Capability Strengthening area have been expanded to provide for collaboration between supported institutions so as to achieve a network of research and research-training institutions in countries where the diseases are endemic. This wider perspective has spawned several additional strategies now being implemented. Among these are:

Training workshops in research management for those scientists playing a managerial role in institutional development programmes. The first of such workshops was held on a global level and plans have been made to hold others at a regional level.

Meetings for scientists working in developing countries for the purpose of exchanging information which has not yet been formally published and for exchanging experiences in institutional development.

Further promotion of training and the development of training programmes in endemic countries.

Promotion of the development of research manpower in institutions on the basis of nationally approved, explicit long-term plans for the development of the institution.

The scientific networks established in the Special Programme have proven to be strong and effective mechanisms both for strengthening institutions and for supporting research in tropical diseases.

24

The Wellcome Trust Network

Peter Williams

The Wellcome Trust very much appreciates being described as having a network of parasitological research. We are bound to say, however, that we do not equate tropical medicine with parasitology nor did we have it in mind to create a network when we started building up our programme in tropical medicine. We take our basis of tropical medicine as the scientific study of diseases that occur in the tropics. In particular, we feel that there are two new major opportunities that have occurred in tropical medicine in recent years. First, the application of modern methodology to basic problems as is so well illustrated by the interests developed through the Rockefeller programme, and second, the application of modern science to the clinical investigation of patients. Our philosophy is that we should support individuals of appropriate quality to undertake the work in which they are interested. We are anxious to encourage them to undertake work on the problems as they occur in man in endemic regions and to maintain their links with home bases in the North. While this is our philosophy, I must admit that we do not adhere to it rigidly and we are expedient according to the circumstances.

The Trust has been supporting tropical medicine research for many years, and even before it was created in 1936 Sir Henry Wellcome had supported work in this field. The classical studies of Balfour and Wenyon in the Sudan and the subsequent establishment of the Wellcome Bureau of Tropical Medicine, which worked on yellow fever, the mycoses, and schistosomiasis, must not be forgotten. Nor must we forget the establishment of the Wellcome Museum of Medical Science, which had a great influence on the education of students who came to London for their training.

My description of our network which, as I have said, is not limited to parasitology, will include some parts that no longer exist. I do this to illustrate that the role of foundations is to create, not a permanent establishment, but a mobile changing intervention that suits the time and place.

Some places will be successful and merit continued support, others will have a limited life and either come to an end or be taken over by someone else who can see their merit once it has been demonstrated. Some of the network, as I have said, would not come under the definition of parasitology, but in fact if you consider the treatment of anaemia or the genesis of malabsorption or the role of trace elements in malnutrition, you inevitably find that there is an overlap, and so you illuminate parasitology indirectly.

Our oldest overseas activity was in Salonika where Foy and Kondi studied malaria after the First World War. They went on to look at anaemia in the tropics, especially in relation to hookworm in Assam and the Seychelles and later in Kenya. The unit they created in Kenya still exists and has moved through megaloblastic anaemia to schistosome immunity and especially the work of Butterworth on the eosinophil to its present interest in schistosomiasis and the vaccine Webbe hopes to develop. It is, however, looking now more towards clinical aspects of tropical medicine. Our overall Kenyan activity includes studies on the evolution of hypertension and the nature of nephritis, vaccination against hepatitis B infection, and the cardiomyopathy syndrome. Our longstanding link with Kenya has enabled us to work on a number of different problems.

The Vellore unit, established some 25 years ago, is now directed by Professor V. I. Mathan. It has always been interested in sprue and the gastrointestinal tract and megaloblastic anaemia. It is well linked to the CRC and other centres in London for virological and biochemical studies. It took part in an international geographical study of sprue some years ago which included comparative investigations in Vellore, Nairobi, Singapore, Haiti and Puerto Rico, and London. That cooperative study was a model for comparative geographical medicine, but we still struggle to find the cause of sprue.

In Bangkok we have a unit interested in cerebral malaria, snakebite, and rabies. The Warrells' recent studies on cerebral malaria have clarified the treatment and opened up the subject. That unit is linked to Oxford, from which studies on thalassaemia and malaria are also undertaken all over the world. The Rockefeller Foundation has recently contributed to that programme.

In New Guinea we have a lecturer based in Liverpool who works on anaemia and its treatment.

In Belem, our unit working on leishmaniasis is now some 15 years old and still doing an excellent job elucidating the biology of the parasite, its vectors, and its reservoirs. The workers there have introduced new techniques for identification of parasites and vectors and have links with London, Harvard, and so on.

Also in Belem, Michael Miles works on Chagas'. He first started work in Bahia as part of the London/Harvard Scheme.

The leishmaniasis programme used to be part of a larger network with a group under Bray in Ethiopia.

Also in the Western hemisphere, we must refer to the links created by the Trust to support the Harvard group at work in Bahia on Chagas' and the work on sickle cell anaemia in Jamaica under Graham Sergeant, which later led to the establishment of an MRC unit as well as the Trust support for the former MRC Epidemiology Unit and the Tropical Metabolism Unit. The latter we are now giving more support to with a group under the leadership of Mike Golden looking at trace elements in malnutrition.

There is also a lot of work in London, and this has various links overseas—Chagas' mega syndrome to Hammersmith, Gambia malaria to Oxford, Sudan nimitti asthma to London, Hammersmith and India on endomyocardial fibrosis and the role of the eosinophil associated with filarial infection.

Much of the network includes modern methodology and so there is a network between the basic scientists and the parasitologists.

Our annual expenditure on the present programme is approximately $4 million. We are very glad to have links with WHO and Rockefeller, the development of whose programmes we believe we have played some part in encouraging.

25

The Great Neglected Diseases: A Global Network for Biomedical Research

Kenneth S. Warren

Ethically, parasitic diseases are not a parochial problem; they are a truly catholic problem. It is not your problem; it is our problem! Chagas' disease is not owned by the Brazilians; African sleeping sickness is not owned by the Nigerians; opisthorchiasis is not owned by the Russians and the Thais. They are problems of global concern to all physicians and parasitologists. We must make maximal use of the world's human and technical resources and work together to solve these problems, using the science and technology of the developed world combined with clinical and field investigation in the developing world in a true and effective partnership.

The ethics of this situation for all parasitic diseases was considered in a paper I recently presented entitled "The Relevance of Schistosomiasis"[1] and I would like to present some quotations from it.

> The first question that arises is whether Boston, Massachusetts, USA has any concern for an infectious disease that affects 200,000,000 persons, most of them living in the rural tropics, and is spreading as the Agency for International Development and the World Bank support agricultural development through dams and irrigation. Today we are constantly aware of, and often irritated by, difficult ethical questions concerning the individual patient: euthanasia, abortion, and medical experimentation are all potential sins of commission. A new kind of ethical question is now being considered, and that involves the health of populations. Should we be concerned with the vast problems of the masses of people of the developing world, in terms of research or in terms of the application of therapeutic or preventive measures? This question is related to potential sins of omission.
>
> My own position in this matter has been particularly well expressed by the philosopher, Peter Singer: "I begin with the assumption that suffering and death from lack of food, shelter and medical care are bad.

My next point is this: If it is in our power to prevent something bad from happening, without thereby sacrificing anything of comparable moral importance, we are morally obliged to do it." But this obligation "requires us only to prevent what is bad, and not to promote what is good, and it requires this of us only when we can do it without sacrificing anything that is, from the moral point of view, comparably important." Finally, Singer notes that this principle takes "no account of proximity or distance. It makes no moral difference whether the person I can help is a neighbor's child ten yards from me or a Bengali whose name I shall never know, ten thousand miles away."

With respect to research on problems like schistosomiasis, a crucial question must be faced: If the scientists of the developed world with their advanced technology, scientific infrastructure, and educational systems do not investigate these problems, who will? Anyone who has experience with the biomedical laboratories of South America, Africa, and Asia is well aware of their dearth of equipment, supplies, manpower, and funding, although there are a few exceptions. Paradoxically, these exceptional laboratories are usually devoted to studying the major health problems of the developed world, such as heart disease (e.g., the great heart institute of Manila) and cancer (e.g., the massive cancer institute of New Delhi). There is, in effect, little or no infrastructure in the developing world for the complex functions of scientific institutions.

Although attempts are being made by major agencies to strengthen educational and scientific institutions in the tropics, the direct involvement of the scientific establishments of the developed world is essential for reasonable progress. This belief is well expressed in a recent statement by the World Health Organization: "Yet it is important that a part of the massive resources for basic biomedical research available in these developed countries be mobilized for work on tropical parasitic diseases. An appropriate solution would be for such institutions to be established and strengthened from national sources, e.g., government or philanthropic foundations." In fact, given the present state of knowledge and methodology in infectious diseases, it seems that application of the scientific power of Europe, America, Australia, Japan, and even advanced developing countries such as Mexico, Brazil, and Thailand will result in rapid advances indeed.

Rockefeller Foundation Program

In December 1977 the Rockefeller Foundation decided to start a new program called the Great Neglected Diseases of Mankind. These diseases involve hundreds of millions of people but are essentially neglected by the great scientific establishment of the developed world. Furthermore, when the program began the funding was exceedingly low, as shown in Table 25-1—particularly so relative to the enormous amounts of money

TABLE 25-1. The prevalence of several of the "great neglected diseases" and their research funding as compared to cancer

	Prevalence: no. of people (millions)	Funding: no. of dollars (millions)
Cancer	10	815*
Schistosomiasis	200	3
Malaria	300	5
Filariasis	300	<1
Amebiasis	400	<1
Ascariasis	1,000	<1

*U.S. Government only

put into research on cancer. Even though funding has been increased in the last five years by initiatives such as the Edna McConnell Clark Foundation's research program on schistosomiasis and the major Tropical Disease Research Programme of the World Health Organization, the total amount of funds is startlingly meager. The Great Neglected Diseases are largely parasitic infections, but also include bacterial and viral diseases such as the diarrheas, dengue, hemorrhagic fever, and measles. The parasitic diseases are a particular problem, however, because tools for their control are in a worse state than those for major viral and bacterial diseases with vaccines available for many of the former and antibiotics for the latter. As seen in Table 25-2, there are no vaccines for any of the major parasitic diseases of man, only one prophylactic drug (to which resistance is developing rapidly in many parts of the world), and therapeutic drugs which are inadequate on the basis of poor effectiveness, high toxicity, and in some cases, putative carcinogenicity.

In December 1977 a proposal was presented to the Board of Trustees of the Rockefeller Foundation "to create a network of high quality investigators who would constitute a critical mass in this field, attract the brightest students, and conduct research of excellence. A significant part of the investigator's effort would be spent in applied collaborative research in developing countries." It was planned to establish units in the mainstream of modern medicine, that is, clinical departments with the highest level of clinical investigation and in great scientific research institutes. There were four basic tenets that were applied in this development:

1. That research would range from the basic level in highly sophisticated laboratories through clinical investigation and field epidemiology.

TABLE 25-2. The plethora of parasitic diseases and the dearth of tools for prophylaxis, treatment, and control

	Infections (thousands of persons infected/yr.)	Deaths (thousands/yr.)	Disease (thousands of cases/yr.)	Diagnostic tests (all parasitic)	Vaccines	Prophylaxis	Treatment
Malaria	800,000	1200	150,000 Fever, coma	Blood	—	± Resistance	± Resistance
Schistosomiasis	200,000	500–1000	20,000 Liver and urinary tract fibrosis	Feces	—	—	+
Hookworm	900,000	50–60	1500 Anemia	Feces	—	—	± Reinfection
South American trypanosomiasis	12,000	60	1200 Heart disease	Xenodiagnosis (kissing bug)	—	—	—
Onchocerciasis	30,000	20–50	200–500 Blindness	Skin snips	—	—	—

Amebiasis	400,000	30	1500 Dysentery, liver abscess	Feces	—	—	± ?Carcinogenic
Ascariasis	1,000,000	20	1000 Intestinal obstruction	Feces	—	—	± Reinfection
Leishmaniasis	12,000	5	12,000 Sore, kala-azar	Splenic aspiration	—	—	± Toxic
African trypanosomiasis	1,000	5	10 Sleeping sickness	Blood	—	—	± Toxic
Trichuriasis	500,000	Low	100 Intestinal disease	Feces	—	—	± Reinfection
Filariasis	250,000	Low	2000–3000 Elephantiasis	Blood	—	—	± ?Carcinogenic
Giardiasis	200,000	Very low	500 Diarrhea	Feces	—	—	± ?Carcinogenic

The plus-minus means that the treatment is indefinite, according to the term, resistance, reinfection, toxic, etc.

2. That it would be investigator-initiated in terms of the problems worked on, which could be any of the Great Neglected Diseases.
3. That support would be long term, probably for at least eight years, and would be flexible although emphasis would be on the development of young investigators and on overseas collaborative research.
4. That the units would be gathered into a global network for communication and collaboration fostered by an annual meeting.

When negotiations began, most institutions had a tendency to spread the funding widely among many individuals working on any problems that might conceivably be considered Great Neglected Diseases. It was necessary to insist on the establishment of a critical mass of investigators into units that would be compact and identifiable. It then became important to develop a critical mass of institutions as rapidly as possible so that young investigators within each unit could perceive significant career opportunities outside of it. Thus, in January 1978, following negotiations of more than six months, seven units were started simultaneously. By the end of the year three more units were added; this was followed by the addition of two units in 1979 and two further units in 1980 to complete a global network of 14 units. In summary, ten of 14 units were begun in the first year and the entire network was established within two years.

Eleven of the units are in the so-called developed world and three in the developing world: one in Latin America, one in Africa, and one in Asia. Five of the units are in departments of medicine and one in a research institute that is largely clinically oriented. These include the Divisions of Geographic Medicine in the Schools of Medicine of the University of Washington in Seattle, Case Western Reserve in Cleveland, Tufts in Boston, Virginia in Charlottesville, and Oxford in England. The sixth unit is the Biomedical Institute for Research in Infectious Diseases in Cairo, Egypt. There are four units devoted to immunoparasitology at Harvard in Boston, in Sweden jointly at the Universities of Uppsala and Stockholm, at the Weizmann Institute in Rehovoth, Israel, and at the Walter and Eliza Hall Institute in Melbourne, Australia. Four biochemistry/pharmacology units are established at Case Western Reserve University in Cleveland, at the Rockefeller University in New York, at the Polytechnic Institute in Mexico City, and at Mahidol University in Bangkok, Thailand.

The units have grown in numbers of investigators from 38 in 1978 to 143 in 1981, and the number of trainees has gone from 17 in 1978 to 80 in 1981. Whereas there were no papers published or in press that were attributable to the units in 1978, the numbers rose from 109 in 1979 to 173 in 1980 to 203 in 1981. Besides a remarkable degree of intra-network collaboration largely brought about through the annual meeting, the GND units are now actively working with 22 less developed countries: Brazil (five units); Egypt, Kenya, and Mexico (four); the Philippines (three); Thailand (two); plus Guatemala, Jamaica, Colombia, Bolivia, Saudi Arabia, Liberia, the

Gambia, Nigeria, Papua/New Guinea, Malaysia, Indonesia, India, China, Singapore, Sri Lanka, and Bangladesh.

Career Development

As an important supplement to the Great Neglected Diseases network, the Rockefeller Foundation began, late in 1978, a career development program for young investigators which provides salary and a significant amount of research support for a period of five years. We were searching for young investigators usually between the ages of 30 and 35 who had demonstrated an interest in the Great Neglected Diseases. Furthermore, because of their energy and determination they had already established a significant record of accomplishment, but had not achieved the stability of major recognition by the granting agencies. The recipients could be M.D.s or Ph.D.s and there were no barriers to any nation or any other environmental or genetic factors. The eight awardees include a Sri Lankan Ph.D. whose major area of interest is malaria, an American Ph.D. whose major interest is schistosomiasis, a Brazilian M.D. Ph.D. working on Chagas' disease, a Canadian Ph.D. working on intestinal helminths, a German M.D. involved in malaria research, an American M.D. working on filariasis and schistosomiasis, another American M.D. working on leishmaniasis, and a French Ph.D. working on schistosomiasis. Further supplementing the program are a series of small grants-in-aid, examples of which are a Brazilian biochemist working in Rio de Janeiro on Chagas' disease, another Brazilian, immunologist at Stanford University also working on Chagas' disease, an Australian clinical investigator working on strongyloides, an American–Venezuelan collaboration on the immunology of leprosy, an American neurologist working on measles encephalitis in Peru, and a Peruvian biochemist working jointly between Peru and Baltimore.

A high degree of coordination has been developed among the lamentably few organizations funding biomedical research on the diseases of the developing world; these include the Wellcome Trust of England, the Edna McConnell Clark Foundation of New York, and the Tropical Diseases Research Programme of the World Health Organization. The Rockefeller Foundation has been on the Joint Coordinating Board of the latter great effort, and the director of the TDR Programme is on the Advisory Committee of the Rockefeller Foundation GND Program.

The mobilization of the scientific power of the world to combat the Great Neglected Diseases by these few organizations in the last five years is already having an enormous effect. This is particularly true because of the application of recent startling advances in techniques and technology which include monoclonal antibodies, genetic engineering, and the use of instrumentation such as high performance liquid chromatography, nuclear magnetic resonance spectrometry, and fluorescence-activated cell sorters. The future looks bright indeed for the achievement of the goals of these programs, which are to develop new tools and apply them for the control of the Great Neglected Diseases.

I find it difficult to end any statement on the Great Neglected Diseases Program without quoting the words written in 1923 by George Vincent, President of the Rockefeller Foundation:

> All that the Foundation is attempting to do, with its relatively limited resources, is to help establish a common front against disease, drawing on the resources and talents of all countries. Whether it is malaria or cholera or plague or tuberculosis or whatever the disease may be, the nations of the world face these enemies of mankind, not as isolated groups behind boundary lines, but as members of the human race pro-jected suddenly into frightening propinquity.

Reference

1. Warren KS (1980) The relevance of schistosomiasis. N Engl J Med 303: 203–206

Discussion

BOOTH: What is the total budget in U.S. dollars?

WARREN: The total budget for the Great Neglected Diseases Program is $1.8 million a year for 14 units. Our contribution is between $50,000 and $150,000 for each unit each year. That is only a small part of their budget; they have been able to obtain support from many other sources.

BOOTH: Is it a kind of dual support system, where you are providing the senior backup and the units are putting in the basic laboratories?

WARREN: Many of the groups are so powerful they are able to command grants from many other sources. Our support to the Walter and Eliza Hall Institute amounts to $150,000, for example; its total budget is $750,000. Our support for the unit in Cleveland is now $50,000 a year; its annual research support totals $1 million. That is the overall pattern. We are aver-aging about $300,000 annually for the Career Development Awards. All told, our Great Neglected Diseases Program is about $2.4 million a year.

CAPRON: We have the essential problem of an international organization and its links with national efforts in research. That is a crucial point if one wants to increase the research potential in many countries.

BOOTH: One of the major awards was for work on Chagas' disease in Brazil in relation to virology.

WILLIAMS: I meant to mention that. During my time in Hammersmith, their main program was on the vasoactive peptides. It was suggested that the syndrome in Chagas' disease might be due to damage to these trans-mitters. Their work stressed that this is the case. That is very appropriate because it is the kind of work that links the two that are part of tropical disease with a technique that was available for use in that situation. It occurred because Pollak came from Argentina.

26

The Funding of
Parasitological Research

Peter Williams

If you are expecting me to tell you where to get funds to support your pet project in parasitology and especially who will look after the new parasitology, I am afraid you are in for a disappointment. In any case Joe Cooke's analysis published in *Current Status and Future of Parasitology* gives a very recent picture. What I will attempt to do is to analyse the sources of funds that are or ought to be available and suggest how they have to be approached. There is not a lot of point knowing how much money is available in each kitty unless you know how to get it out. To get it out requires careful homework, not only about the motives of the fund holder, but about the objectives of the person who wants the money.

Money for research is provided for a variety of reasons which may be classified as:

Commerical: to yield a product for profit;
Academic: to assist training, to extend knowledge for its own sake;
Humanitarian: to help poor and deprived people
Political: to influence foreign policy, in connection with defence, to safe-
    guard the health of others.

The sources of funds mirror these motives and are geared to them. Essentially they can be divided into (1) industry, (2) government, (3) private foundation and Individuals Charity (we call all this charity).

If I assume that the objective of this volume is to increase activity in parasitological research and that therefore more funds should be attracted towards this field, then the next step is to decide what is the motive for doing this. Is it to improve the health of people in the tropics, to advance academic knowledge for its own sake, or to safeguard the health of the armed forces of the United Kingdom in Zimbabwe and Honduras or the United States in Vietnam and El Salvador?

281

It is essential to know the purpose if you wish to improve the funding. It then becomes obvious who should be approached. Thus it is not much use going to industry if the work is very theoretical, and the charities are hardly likely to support research undertaken for defence purposes. But to every one of those general rules there is an exception. Commerce may take a long view and support theoretical research of remote practical relevance and charities may also be less directly concerned with the immediate humanitarian aspect, but these variations are a matter of intimate knowledge of the particular organisations from which one wishes to receive a grant. But even then it is not sufficient to read their literature, because what really matters is the individual or group of individuals who make the policy and choose the topics to support. Thus, for example, foundations operate very differently. You have only to contrast four foundations that have made major grants to parasitological research in recent years: The Rockefeller set out under the dominance of one man with a zeal for his subject to recruit the best people he could find to the field from his intimate knowledge of the subject. The Wellcome is a more responsive organisation taking a more general view of tropical medicine and using the broad advisory system of evaluation of people. The Wolfson identified the special problem of tropical medicine in Britain after the publication of the *Transactions* supplement and offered a large sum to the two schools of tropical medicine in England to encourage them to be more interrelated and more associated with other institutions. The Clarke with its one disease concept aimed at a practical end point, and so on. The person who wants money has to consider these points and frame his proposals to meet them. Does all this suggest a simple con game—frame your application to get the grant—or a genuine shaping of the direction of research by the funding organisation? Here we meet a fascinating contrast of views between those who believe that it is only the scientist who can plan research and the organisation that says it knows where it wants to go and it will attract the people it needs.

I would like to suggest that these two standpoints are coming together quite rapidly so that there is no need to have the antagonism between scientists and administrators that has been there in the past. Research workers will not be so wary of practical suggestions if they come from members of their own group who they respect, and in this the policy of the Wellcome Trust to recruit able scientists to take charge of their programmes is an example.

I believe that in seeking funding for research the principle problem is for the research worker to take a good look at his personal motives and then fashion his programme towards achieving these ends in partnership with a funding body that can understand his purpose. For parasitology there are special funds that are not available to others and those who want support for this field have to see that they identify their needs in the right quarter.

All this is I suppose fairly obvious, but it is my experience that people undertaking research do not usually take the trouble to understand the motives of those who finance it and as a consequence waste a lot of time applying to the wrong organisations and becoming frustrated by the ogre they call administration.

So much for the funds that are available. But of course what is suggested by this volume is that parasitology research is underfunded and that our real task is to persuade the present organisations and others not already in the field that more funds should be channelled in this direction. What therefore is the strength of the case? What case is there for saying that parasitology matters more than the world recognizes? Since we will still be dealing with the same sources of funds, what message can we give to them to increase their financial interest? I can see that the recent application of modern science to parasites provides an interesting new world for study, and so the academic case is strong. I know that parasites are a major cause of a lot of suffering among the poor. It is therefore possible to persuade private organisations with the aim of helping the advancement of knowledge or helping the suffering poor that they should give more help. But can we marshal strong enough arguments to persuade industry and government to invest large sums in parasitology? They will wish to see a product and a profit from their enterprise. Unless you can formulate a plan in which industry and government can see a way to make a major impact on health through the results of the research you want supported. I'm afraid they will regard you as simply another lobby of relatively poor weight.

This volume has related the present state of parasitology and the last two sections (Chapters 24, 25) have suggested how the subject can be expected to advance. What is now needed is conversion of this information into a strategic plan directed at the potential sources of funds. I think that the most useful way to discuss this paper would be to try to decide where the future of parasitology lies and where the funds should come from to take it there.

Discussion

HARINASUTA: As the only participant from a developing country, I would like to comment on future work on parasitic diseases, which are a major public health problem in Thailand, especially in relation to the international network that has been mentioned. The international network includes "Type One" institutions in the developing countries; that is what we call the Tropical Medicine, which means that all institutions that do research on parasitology meet once a year.

"Type Two" are institutions in developing countries. The Southeast Asia Ministers of Education Organization (SEAMEO) receives assistance from the developed countries, including Australia, France, and New Zealand.

Then we have TROPMED, tropical medicine in public health, for which we also need assistance from the developed countries.

Through the WHO-TDR Program, Lucas has supported us over the last three or four years. The assistance from the developed countries comes in the form of consultants, equipment, and supplies, staff fellowships, development and funds for research projects, and funds for training courses.

The third type involves cooperation between institutions in the developed and the developing countries; so far this is the only cooperative program on specific diseases. The SEAMEO and TROPMED projects were established in 1967 for education, training, and research in tropical medicine and public health in Southeast Asia. Member countries at the present time include Indonesia, Malaysia, the Philippines, and Thailand.

The ministers of education are currently concentrating more on training courses. We have 11 courses in such locations as Bangkok, Kuala Lumpur, and Manila. Six concern parasites, and four of them are in Kuala Lumpur, Djarkarta, Bangkok, and Manila. They are plotting newer directions for parasitology such as urban health and occupational health.

The TROPMED program also works on problems connected with parasitology.

With regard to research development in Southeast Asia, the epidemiology of tropical diseases in relation to public health has a high priority. Warren has said that epidemiology is most important in our country because before we can control diseases we have to know more about epidemiology. With regard to the control of tropical disease, we have been conducting training courses in mosquito genetics over the last two years. We are now working on the immunology of parasitic diseases, which we understand is very difficult.

SOPRUNOV: Let me say something about the possibilities that can be found in the Soviet Union and what we could do to help in this field. As far as I am concerned, I would do my best to develop contacts and to establish scientific links with tropical institutes and research laboratories abroad. Our country has a high scientific potential that can be used in this field.

There is another thing I would like to draw your attention to: the so-called boomerang effect. Let us say ten scientific institutes are receiving support. If a scientist in one institution publishes a paper on some new and interesting work using a new technique such as liquid chromatography, I am astonished to find several of the other institutions order this equipment. It's the same in our country: a lot of institutions spend money on the same equipment used in the other laboratories. We are wasting a great deal of money because of this duplication of equipment. Something must be done to stop this boomerang effect. Perhaps each laboratory should be specialized, so they are not all doing the same kind of research.

MAO: I protest that Harinasuta did not include the People's Republic of China among the developing countries, for it is clearly one. We are economically and technically behind the needs of our people. It is true that China was cut off from the outside world for nearly a quarter of a century, but we now want to contribute.

WARREN: I believe Mexico could also be considered a developing country.

MARTINEZ-PALOMO: We are spending about $80 million a year just for malaria control programs, yet not a single laboratory in Mexico is working on research in malaria. Obviously that means such funds can very easily go to support other types of research.

WARREN: We have heard an elegant talk by a great donor. I would like to hear how recipients feel.

I am in a somewhat different position than Williams because I was a recipient for so many years. I can assure you, referring to the title of Allen Gregg's biography *The Difficult Art of Giving,* it truly is difficult. But it is a different kind of difficulty, and one that rankles people to a large extent who have to ask for funds. We recognize that, and we are very concerned about it.

The Rockefeller Foundation has tried as much as possible to make it a partnership. When we start a program, it is a completely open relationship between donor and donee, except we have to say "no" very often, and that hurts on both sides.

MITCHELL: The one issue we have not dealt with is that when foundations infuse a particular field should they feel an obligation to keep picking up the tab for the people they support? If a foundation puts money into a program to train young people, and they get interested and enthusiastic, this automatically means an ongoing commitment to keep these people in business. That to me is unfortunate. I feel, on the contrary, that a foundation should induce competent people to do good work with generous funding, and they should then be in a strong position to go to their own national funding organizations with results in hand, based on support a Rockefeller or a Wellcome has given them over the years.

WILLIAMS: I didn't intend to imply that. I mean to say that the length of time we support a program has to be judged according to whether it is being taken up by other sources. If we fund a project that for some economic reason or other isn't picked up, we shouldn't drop it simply to do something else. I of course believe it is necessary to have funding to start a program going. I am totally with you about that.

WARREN: That is absolutely correct. One element that hasn't been mentioned here is merely a circumstance of history. The whole world is going through a recession at this moment, and to my mind, knowing the quality of work going on in the Wellcome unit, the TDR program, and the Rockefeller unit, if the global situation were a lot easier there is no question that most of this work would be picked up, and with ease.

ARNON: I think your comment is correct, Dr. Mitchell, that we should be able to apply to our national governments for more money, but in some places it is simply not available.

In Israel, for example, the priorities are such that tropical medicine is not high on the list. On the other hand, the amount of money needed is substantial. I personally will do what I find exciting to do whether or not

I am supported. If I don't have money, I will somehow get temporary support from other sources. In general, however, if some foundations give money for earmarked purposes I shall continue and not depend on other countries.

MARTINEZ-PALOMO: For us the most important thing is that we have been treated as equals by Warren and by the network we are affiliated with. It is very important for us to have recognition from outside, for it led us to the point where our unit could become independent from other departments.

It also facilitated cooperation with other laboratories inside as well as outside Mexico. This has given us a unique opportunity to learn from our colleagues and to improve the level of training of our staff.

Also—and this is very important—we have to strive to reach the same level as the excellent units that form the network. We are particularly grateful to Warren because from the day Williams suggested I take a plane to New York to ask him for money, Warren has treated me as an equal.

WARREN: That was such a nice thing to say that I hate to call on Nelson to have the last word and balance it off.

NELSON: In all of this discussion we have forgotten the silent majority. There is a crowd of characters sitting out in the wilderness doing research on vectors of onchocerciasis. I recently returned from India, where some superb field work is going on that is totally unrecognized by anyone because it is a national research program.

They are not asking for support from the TDR or from Wellcome or from the MRC. It is a program of Indian National Government. Some of the work is of very high quality, based somewhat on the old biology. I hope that, apart from the network of modern biology laboratories that this group ensures in the interest of its own future, it will collaborate with this "silent majority" in the field.

WARREN: I would like to thank you. I believe this is one of the most exciting meetings I have ever attended on every level, from the scientific through the social. I can't thank you enough for a highly stimulating four days.

Index